Eloquent Animals

A Study in Animal Communication

by Flora Davis

INSIDE INTUITION

ELOQUENT ANIMALS

ELOQUENT ANIMALS

A STUDY IN ANIMAL COMMUNICATION

How Chimps Lie, Whales Sing, and Slime Molds Pass the Message Along

Flora Davis

Coward, McCann & Geoghegan, Inc.
New York

Library of Congress Cataloging in Publication Data

Davis, Flora.
Eloquent animals.

Bibliography: p.
Includes index.
1. Animal communication. I. Title.
QL776.D38 1978 591.5'9 77-18021
ISBN 0-698-10892-2

Printed in the United States of America

For my father

Acknowledgments

This book wouldn't have been possible without the generous help of scientists who are studying animal communication. They told me where to find material, they patiently answered questions, and eventually they read over parts of the completed manuscript and let me have their comments on it. Many of them also spent considerable time introducing me to their particular animals and explaining how they study them.

I owe particular thanks, then, to: Professor George Barlow, Professor Colin Beer, Professor Norman Bleicher, Professor John Bonner, Professor Robert Capranica, Dr. William Cummings, Professor Roger Fouts, Dr. Michael Fox, Professors Allen and Beatrice Gardner, Professor James Gould, Professor Donald Griffin, Professor Ronald Hoy, Ms. Ellen Kwait, Professor Donald Kroodsma, Professor Erich Klinghammer, Mr. Terry Lim, Mr. Paul Loiselle, Professor Peter Marler, Professor Emil Menzel, Ms. Penny Patterson, Professor David Premack, Professor Duane Rumbaugh, Dr. Sue Savage, Professor Herbert Terrace, Dr. Howard Topoff, and Ms. Marianne Yeutter. I also want to thank Professor Theodore Meth, whose expertise is in the field of animal rights, and some old friends who research *human* communication: Professor Ray Birdwhistell, Dr. William Condon, and Dr. Adam Kendon.

I'm grateful, too, to Ms. Margaret Borgstrand, who was so patient while I struggled to learn sign language, and to my editor, Ms. Patricia Soliman; her feeling for the direction the book should move in was at times considerably better than mine. Ms. Janice Harayda read over parts of the manuscript for me, and Dr. Mike Tayyabkhan lived through it with me, chapter by chapter. Lastly, I'm grateful to my children, Jeffrey and Rebecca Davis, who sat through chimp sagas and slime-mold stories alike with gratifying enthusiasm.

Contents

1.

Conversation with an Ape

"Ape's brain," Zira concluded, "has developed, is complex and organized, whereas man's has hardly undergone any transformation."

"And why, Zira, has the simian brain developed in this way?"

Language had undoubtedly been an essential factor. But why did apes talk and not men? Scientific opinion differed on this point. There were some who saw in it a mysterious divine intervention. Others maintained that ape's mind was primarily the result of the fact that he had four agile hands.

"With only two hands, each with short, clumsy fingers," said Zira, "man is probably handicapped at birth. . . ."

—PLANET OF THE APES,
by Pierre Boulle

She was shaggy and brown and stood about waist-high. Her body was all animal—she was a four-year-old gorilla—but the intelligence in the steady brown eyes was somehow disconcerting. She looked me over thoughtfully, her face close up against the wire mesh that separated us. After a minute or two, she went and got a stool and climbed up on it so that we could be at eye level.

Her name was Koko and she lived on the Stanford University campus, sharing a five-room trailer with her trainer, Francine (Penny) Patterson, who is teaching her sign language. On the day I was there, goats wandered freely in the fenced-in enclosure that surrounded the trailer. Inside, Koko had the run of the living room and kitchen, but a wall of wire mesh and a securely locked door kept her from joining me on the doorstep or the animals in the yard.

This restriction apparently was not altogether to her liking, for Koko soon got down off the stool and went over to Penny, who was in the room with her. Taking her by the hand, Koko led her to the outside door and began to make rapid gesticulations with her hands. Although I had acquired a smattering of American Sign Language, the movements were much too fast for me to follow, so Penny translated for me:

"Koko is saying, 'Out me please key open.' "

Penny signed back a quiet no.

Koko stood still, considering the situation. Then, like a mischievous child, she suddenly raced through the kitchen, leaped onto one of the counters, charged across the top of the stove to the other counter, where Penny's typewriter was sitting open, gave the typewriter a hearty thump, and leaped down to the floor again.

There was in that sudden charge such a sense of unleashed power that it set me back on my heels. I peered apprehensively through the mesh at Penny. She is a tall young woman but very slender and obviously no match for an obstreperous gorilla-child. Nevertheless, she seemed totally calm.

"Go to your room," she said firmly to Koko. She said it aloud in English, since Koko understands spoken English as well as sign language.

Head down, Koko knuckled across the floor into the living room and climbed up on the wooden shelf that is her bed. Penny remained in the kitchen and locked the wire-mesh door between the two rooms.

For a while, Koko sat quietly on her bed, chin down, brooding, just as any other chastened four-year-old might. Finally she looked over at Penny and began to sign. Once again Penny translated for me:

"She says, 'You open door out now. I be good, be quiet.' "

Taking Koko at her word, Penny unlocked the door.

* * *

How can I convey what it's like to watch an ape communicate in human terms—in signed words? It is not only startling, it's downright disorienting. Several times that day I caught myself wondering what Koko thought of me—and then realized that this was not a question I had ever expected to ask about an animal. We may be indifferent to animals, amused by them, curious about them, sometimes frightened of them, but when do we ever concern ourselves with their opinions? Later, talking to friends, I made a joke of the confusion I felt, saying that I was not sure whether the problem was that Koko was almost human or that I was almost gorilla. All the same, the feelings were real and disturbing, an odd kind of blurring of my sense of my own identity as a human. I have since learned that the people who work with gorillas and chimps sometimes do have this reaction.

I first became interested in the efforts to teach language to apes because this seemed like a way to learn more about *human* communication, about language itself. In addition, I had written a book about human nonverbal communication and was well aware that, in many ways, it bears a startling resemblance to the necessarily nonverbal exchanges of the other primates. However, in the beginning at least, animals were not my chief interest.

Perhaps it shouldn't come as a surprise to humans to learn that apes can "talk." And yet it does. Language has always been the great barrier, the special ability that set us apart, so that we could logically divide all the earth's creatures into human and nonhuman. To many of us, the fact that we can talk and think in concepts seems to explain quite adequately why we—and not chimps or elephants or killer whales—have built cities and developed the theory of relativity, have walked on the moon. To realize that apes can learn the rudiments of language and perhaps much more is, then, a bewildering and potentially humbling experience.

Yet to be able to communicate with animals has also been a persistent human dream. It is preserved in the mythology of many cultures and must, I think, be an almost universal childhood fantasy. It's a dream that today seems on the verge of coming true for two different reasons: we are building up a store of knowledge about how animals normally communicate with one another—frog to frog and whale to whale; and we are now able to communicate with apes directly.

Already, in the ape language experiments, we have come a long way. Consider these milestones:

- Within the last ten years, well over a dozen chimpanzees have been taught to sign.
- The first chimp-to-chimp signed conversations have taken place.
- At a chimp colony in Oklahoma, animals who have never had lessons in sign language are beginning to pick up signs from chimps who have had lessons. Soon we will know whether a signing chimp mother will spontaneously teach sign language to her baby.
- Nine chimps have learned, not to sign, but to read and write, using specially designed word-symbols rather than the alphabet.
- The work with Koko has demonstrated that a gorilla can learn sign language as readily as a chimp. Preliminary studies suggest that orangutans might also be good students.

Lest the reader think it's a trivial matter to learn sign language, let me describe my own experience. The accomplishments of the signing apes became very real to me in a personal way when I decided to take a few lessons in American Sign Language myself. I had some idea, I think, that if I could just zip through a crash course in ASL, I could hold up my end of a conversation with Koko or one of the signing chimps. Although I was never exactly adept at languages, I was sure a gestural language would be different: I expected something simple, with lots of pantomime and body English thrown in. However, I soon discovered that, while there were elements of pantomime, ASL is a language as complex as any other. My teacher's hands, flowing through a signed sentence, were as hard to follow as a comment rattled off in Greek. I found myself desperately rummaging through my memory for vocabulary: surely that last gesture was familiar—one of those I memorized only last week. . . . Though I persevered through half a dozen lessons, when I was introduced to Koko, it took me only a minute or two to realize that she was far, far ahead of me in signing, in fact, completely out of my league.

When an ape learns to sign, then, it is no small accomplishment; nor is the ability to read easily achieved. Once I realized these things, my questions began to pile up.

Why is it that apes can learn to sign and to read, and since they *can* learn to sign, why haven't they been communicating by gesture all along? Or have they? Is it possible that if we just knew what to look for, we would discover that they already exchange information to a much greater extent than we ever imagined?

Is human conversation really totally different from animal communication, or is it just different in degree, in complexity? How, in fact, did human language ever evolve?

Are there other animals that can learn language? How much do dogs understand, for example, and how much can they be taught? Is it really possible that dolphins "talk" in the equivalent of words? Is it true that bees are better at exchanging information than most of the higher animals; that bee communication is, in fact, more like language than is the signal system of any other species?

In search of a broader perspective, I began to interview both the people who are doing research on apes and language and those who are working on garden-variety animal communication—on the signals that birds, bees, and whales exchange. I soon gave up any attempt to be comprehensive—there was no way I could talk with everyone who was doing important research. Instead, within the limits of my travel budget, I tried to hit the highlights in what became a kind of report from the front lines of animal research.

As I got deeper into the project, I discovered that I had embarked on an unforgettable adventure. There was the day I spent at Princeton University, where in the morning I was treated to a bee's-eye view of what really goes on in a hive, and in the afternoon I found myself peering down a microscope while a biologist explained how slime-mold amoebas signal to one another. There was the time I spent in San Diego, where for two glorious days I trailed a whale expert around, and one afternoon watched with him while killer whales called back and forth. I saw a songbird having a singing lesson at one lab and ants resting up before a slave-taking raid at another. I learned how to mouth-fight with a fish. I was kissed by a gorilla!

At some point, I began to feel that I might be in on the beginning of something big, perhaps a sea change in the way we think about animals. Almost everywhere I went I heard the same story: animal communication is much, much more complex than we ever dreamed it was. Eventually I came to feel that the real language barrier may be the human belief that language is totally different from animal communication—just as humans are totally different from animals.

Over the centuries, the human ego has suffered a number of severe blows. We have learned to live with the knowledge that the earth is not the center of the universe; that humans are related to apes; that we are not the only tool-using—or even tool-making—species. The belief that we are the only animal capable of language will be at least as difficult to give up, since language is tied to our sense of identity as humans.

In the end, I spent two and a half years researching this book (with time out to write magazine articles for the income), and the work was pure joy. It's my hope that reading it may do for the reader what writing it has done for me. The world no longer seems quite the same place, now that I know what the crickets are chirping about and how much information is coded in a bird's song. I have discovered that a lot goes on in my fish tank that I was never aware of before, and some of the things I see dogs and cats do make a new kind of sense to me.

Somewhere along the way, I stopped thinking of evolution as a continuous upward push, with humanity as its greatest achievement. Instead, I began to understand what some of the animal behaviorists mean when they say that evolution is not a vertical but a horizontal process, that the only real criterion is species survival.

Most of all, though, I have come to feel that it's incredibly myopic to live as if humans were the only relevant and interesting species. There are on this planet so many different ways to *be,* to experience life. Animals can see, hear, feel things we can only imagine. They can communicate in ways we can describe but never experience.

This book will take as a starting point the quite remarkable accomplishments of the "talking" apes, and will then go on to examine how other, simpler creatures communicate, beginning with amoebas and returning in the end to apes and the ways in which human interactions resemble those of the other primates. In the final chapter, I will consider what language is, seen against the backdrop of communication in the broadest sense. I'll ask again what language means, what it does for us, what part it plays in being human.

2.

The Washoe Project

I saw the woman just above us, perched on the rocky ledge from which the cascade fell. . . . I thought she was going to speak, to give a shout. I was expecting a cry. I was prepared for the most barbarous language, but not for the strange sounds that came out of her throat; specifically out of her throat, for neither mouth nor tongue played any part in this sort of shrill mewing or whining, which seemed yet again to express the joyful frenzy of an animal. In our zoos, sometimes, young chimpanzees play and wrestle together giving just such little cries.

—PLANET OF THE APES,
by Pierre Boulle

Imagine this:

Swarms of killer bees are moving north from Brazil (so far, the story is true). As they cross the border into Guatemala, there is an unfortunate incident: a farmer is stung to death. A public outcry follows, and the bees are accused of murder. In what is described as the court case of the century, they are tried *in absentia* (no one really wants them in court), convicted, and sentenced to death by extermination.

Absurd? But ancient precedents exist. Throughout much of recorded history, animals were held responsible for their actions, just as people were, and they were often tried for breaking human laws. During the Middle Ages, pigs were hanged for murder; dogs were jailed for assault; insect pests were sometimes excommunicated by the Catholic Church; and in at least one instance, in A.D. 864, a swarm of wild bees *was* convicted of murder and sentenced to death by suffocation.

Today we know that animals don't operate psychologically in the same ways humans do, and the idea of holding them legally responsible for their actions seems ludicrous. However, it's worth noting that the lines between human and nonhuman intelligence were not always so clearly drawn.

Nor has it always seemed obvious that animals haven't the mental capacity for language. As recently as the turn of the century, "talking" horses and pigs regularly made the rounds of country fairs, tapping out messages with their hooves in numerical codes, and many people believed they could actually add, subtract, and spell. Talking dogs were exhibited, too. Most pawed out numbers on the ground, but attempts were also made to train dogs to speak. Alexander Graham Bell once taught a dog to grate out "Ow ah oo gwah mah," which was as close as it could come to "How are you, Grandma?" The dog "talked" by growling steadily while Bell manipulated its throat and jaws.

Eventually most people became convinced that animals are in fact prisoners of their instincts, capable of expressing emotions and intentions nonverbally, but of very little more; yet well into this century, it was thought that apes might prove to be exceptions to the rule.

Humans have, in fact, made a number of attempts to confer language on apes. They have been tutored in English, Russian, German, and undoubtedly in other languages as well—for the most part, unsuccessfully. The record for vocabulary was held by a chimp named Viki, who had learned to say four words by the time she died, at the age of seven, in the early 1950s. Viki was raised in a human home like a human child, and an immense amount of effort went into teaching her those four words. Since she had had every human advantage, her meager achievement convinced most scientists that chimps were simply incapable of language.

However, about ten years ago, it occurred to psychologists Allen and Beatrice Gardner of the University of Nevada that a chimpanzee might be able to learn a gestural language although it couldn't learn a spoken one. The Gardners (they are husband and wife) noted that Viki had accompanied each of her four words with its own particular gesture, so that an observer could tell what she was saying from the movements of her hands alone. They were also aware that chimps in captivity spontaneously use a kind of primitive sign language to communicate with humans—asking for food, for example, with the classic palm-up gesture that human panhandlers also use. Perhaps chimps had failed to talk only because they were physically incapable of uttering words.

And so the Gardners set out to teach American Sign Language to an infant chimpanzee they named Washoe. ASL (it is also sometimes called Ameslan) is the gestural language used by the deaf in America, and it is a true language with a grammar of its own. Each sign represents a word or an idea, and although some signs mimic what they represent, most are not iconic. Since ASL is currently being closely studied by linguists, the Gardners knew that eventually they would be able to compare Washoe's progress in learning the language with the progress of human children who are deaf.

Washoe, born in Africa, was about eleven months old when she arrived in Reno, Nevada, in June, 1966. The Gardners promptly installed her in a trailer in their back yard, where she had her own bedroom, bathroom, and kitchen, sturdily furnished and stocked with clothes, toys, and books. She was very much like a human baby of about the same age: she had only a few teeth, she had just begun to crawl, and she slept a lot of the time. In the wild, chimpanzee infants aren't weaned until they're four or five years old; they reach puberty at seven to ten, and sometimes live well into their forties.

From the beginning, spoken English was never used in front of Washoe, since the Gardners were afraid it would distract her from the signs they wanted her to learn. However, in other ways, they tried hard to reproduce for her the circumstances under which children learn language. Humans were with her all day long, and they chattered to her in sign language much as mothers talk aloud to babies. There were a few problems, however, that mothers don't normally have, for Washoe was so active that she often took no notice of the gesticulations of her human companions, who frequently had to abandon their signing anyway and use their hands to keep her out of mischief. But she often made things easier for them by spontaneously imitating signs. They also used *molding* to teach her: they would take her hands and put them through the correct motions until she caught on.

By the time Washoe had been with the Gardners for seven months, she knew four signs: "come gimme," "more," "up," and "sweet" (the signs for "come" and "give-me" are similar beckoning gestures). "More" is done with hands held in front of the body, by touching the fingertips together several times. Washoe first learned to sign "more" when she wanted more tickling—young chimps love to be tickled—but soon she was also using it when she wanted more of being dragged around the floor in a laundry basket and when she wanted seconds at meals. In other words, she demonstrated the ability basic to learning language of being able to *generalize* from the particular.

The real breakthrough came ten months into the project, when she first combined two signs. "Gimme sweet," she demanded. At the time she was twenty-one months old, close to the age at which human children begin to pair words. The Gardners had not molded or otherwise urged sign combinations, though of course they had signed to her in phrases and sentences themselves. However, Washoe was not simply imitating their combinations, for she soon began to invent some of her own. "Gimme tickle" was her invention, and she called the refrigerator "open food drink," though the Gardners had always referred to it as the "cold box."

Then one red-letter day Washoe invented, not a new phrase, but a whole new sign. Because the Gardners couldn't find an ASL sign for "bib" in their manuals, they had taught her the one for "napkin," which is made by wiping an open hand across the mouth. One evening at dinner, Allen Gardner held up a bib and asked her to name it. She made several wrong answers, then suddenly, using both hands, traced an outline of a bib on her chest. This was such a logical invention that the Gardners briefly considered abandoning the "napkin" sign and adopting Washoe's version. They decided against it because, as they wrote afterward, "The purpose of the project was, after all, to see if Washoe could learn a human system of two-way communication, and not to see if human beings could learn a system devised by an infant chimpanzee." Three months later, while lecturing at a California school for the deaf, they learned that the correct way to sign "bib" in ASL is to outline a bib on the chest!

The Gardners noted that there are two ways to explain Washoe's ability to invent signs: either it is something she was able to do because she had been exposed to sign language, or it's something well within the capabilities of chimps living in the wild. In other words, whether they actually communicate by a gestural language or not, they are capable of inventing one.

As Washoe learned more and more signs, she began using them all day long, making comments, starting conversations. At first, she seemed to assume that every creature she met knew ASL, and she signed away at all humans and even at dogs and cats. The Gardners reported, "New research assistants have commented on the singular humiliation of having Washoe sign to them ever so slowly and carefully when they were only beginners at Ameslan."

Washoe also liked to sign to herself. Leafing through a magazine, she would make a running commentary ("That food. That drink."). Bent on mischief, she would tell herself, "Quiet," as she crept across the yard. Racing to get to her potty chair in time, she shook her hand frantically in the "hurry" sign. Once she was drawing on a piece of paper when she suddenly stopped, told herself "Up hurry," and raced up a nearby ladder. Soon she climbed down again and returned to her drawing—only to sign "more up" and head for the ladder again.

Even as her signing vocabulary grew, Washoe continued to produce the vocalizations and facial expressions that are part of normal chimp communication. On one occasion, this led to a misunderstanding. The incident occured when Washoe was small, on a night when she was running a fever and the Gardners simply couldn't get her into bed. Finally, at two or three in the morning, Beatrice Gardner left the trailer, exhausted, while her husband stayed on. However, with her departure, Washoe became very upset and jumped on a table. She was wearing the chimpanzee fear face, an expression which displays a lot of teeth, but Allen Gardner didn't recognize it. Thinking that perhaps Washoe was out of her head from fever, he was just bracing himself for whatever might come next when—apparently realizing that he didn't understand—she signed, "Come hug." This, perhaps more than anything else, illustrates the part sign language had come to play in her life.

The Gardners considered a sign a permanent part of Washoe's vocabulary when she had used it appropriately and spontaneously at least once a day for fourteen or fifteen consecutive days. In the beginning, they were able to keep fairly complete records of all her signing. When this become impossible because she was signing so much, they instituted twice-a-week tests using slides projected inside a specially designed cabinet. Washoe would open a door to see what was on the screen, and her human companion would then sign to her, asking her what she saw, and record her answer. Young chimps are highly active animals who are easily bored, and the great advantage of the Gardners' test system was that it was self-paced. Washoe could play the game for as long as it interested her, then abandon it to do something else, perhaps returning to it later. In one marathon seven-and-a-half-hour session, she actually identified ninety-nine slides in between meals and play.

By the time Washoe was an obstreperous five-year-old, she knew 160 signs and was stringing them together 4, 5, 6, and more at a time. She had disproved the assumption that language is completely beyond the chimpanzee. She had her critics—linguists who insisted that her signing was too rudimentary to qualify as languge—but still it seemed clear that humans had underestimated chimps.

Anatomists now believe that if chimpanzees have not talked, it is at least partly because, as the Gardners suspected, speech itself is beyond them. The chimp vocal tract is apparently unable to produce human sounds, and there is also a good chance that they can't control their vocalizations the way humans can. Most of the time, chimps are silent. When they do make noise, it's usually because they're excited, and the sounds they make then are fairly predictable, given the particular situation, and go with predictable facial expressions, almost as if both were automatic reactions.

Keeping deliberately silent also seems difficult for a chimp. Viki's human parents reported that at times she used great ingenuity to steal sweets, only to give herself away with grunts of pleasure as she ate. Most species have in their repertoire some responses that are innate, as is the courting song of the cricket, and some that have to be learned. It seems likely that chimp vocalizations are pretty much innate and therefore difficult for the animal to control. On the other hand, chimp gestures and postures express a great deal and don't appear to be prewired in the same way—all of which may explain why Washoe could sign though she couldn't speak.

In many fascinating ways, Washoe's progress as she acquired signs paralleled the language acquisition of human children—which is, in itself, a mysterious and much underrated phenomenon. Babies respond to language from a very early age. An infant only a few months old already babbles in its native language: a Chinese baby makes the sounds appropriate to Chinese; a French baby, those characteristic of French. At the age of about a year, a child begins to utter recognizable words. At first they come singly, as Washoe's early signs did. Then the youngster begins to put words together two at a time, marking them off vocally with normal adult stress and pitch. When a small boy says "Mommy drink," it's normally not hard to tell, just from the way he says the words, whether he is asking a question, making a statement, or issuing an order, and it's also quite clear that the two words go together. In fact, intonation—the rise and fall of the voice—is the chief grammatical device of the very young.

Washoe used the grammatical conventions of sign language to make the same kinds of distinctions. At the end of a pair or a string of signs, her hands relaxed or dropped out of the signing space if she was making a statement or a demand. To indicate a question, she kept her hands in position at the end of the string and made eye contact. When she did this, even people not familiar with ASL often sensed that she was asking a question.

As children learn language, they demonstrate their ability to conceptualize: the three-year-old girl who uses the word "dog" not only for the family poodle but also for the collie she sees on the street obviously has a generalized idea (a concept) of what a dog is. Actually, at the stage where their vocabulary is still very limited, children often overgeneralize. Confronted with something they don't have a name for, they tend to forge ahead and pick from among the words they do know the one that seems to stand for something similar. The little girl may at first use the word "dog" not only for all dogs but also for cats, squirrels, and any other small, furry, four-legged animal.

Washoe proved very early in the project that she, too, worked from concepts. For example, she learned to sign "open" when she wanted a door opened, and went on to demand "open" when she needed help with drawers, jars, and even faucets. Like human children, she also tended to overgeneralize, and when she did, she sometimes provided a fascinating clue to the way her mind worked. She signed "hurt" quite appropriately for scratches or bruises, but she signed "hurt" the first time she saw a human navel as well. She used the sign for "flower" when she was shown a bouquet, but she also used it when she walked into a kitchen filled with the smell of food cooking and when she opened a tobacco pouch. Evidently, to her, the salient feature of a flower was its scent rather than the way it looked, and on that basis, she generalized the sign. It wasn't until the Gardners taught her the correct sign for "smell" that she got the meanings sorted out. Any parent of a two- or three-year-old child has similar stories to tell.

But of course language is not just learning words or even concepts. The essence of language is syntax, the rules by which words are assembled into sentences. A parrot can easily be trained to squawk, "A bucket of blood." The same bird might already know the word "crackers," but it's extremely unlikely that it will ever come out with "A bucket of crackers," since a parrot can't make up new sentences. It can only repeat what it has heard.

No human learns a language by memorizing all of its possible sentences—a hopeless task, anyway—since the human mind doesn't swallow a sentence whole. Instead, we break up what we hear into pieces and analyze the relationship between the pieces—a relationship made clear by syntax. Word order, for example, is one syntactic device that English depends on heavily. To the English-speaking person, it's quite clear that "Paint the box brown" is an entirely different directive than "Box the brown paint." Syntax is also expressed by prefixes, suffixes, verb tenses, and all the other devices we learned to think of in school as "grammar."

All human languages have syntax, and so all languages are incredibly open-ended: from a finite number of words, by endless permutations and combinations, they can create a virtually infinite number of different sentences. There are hundreds of ways to say almost anything. In fact, linguists point out that any time we hear a twenty-word sentence, we are probably hearing that exact combination of words for the first time in our lives, since there are so many, many other ways to convey the same information.

Nobody sets out in an organized way to explain grammar to a two- or three-year-old, and so children must abstract the rules of syntax for themselves, just from listening to adults talk. What is remarkable is that they are able to do this successfully in spite of the fact that the speech they hear is highly imperfect and full of grammatical errors and fragmented sentences, and that they may seldom hear the same sentence spoken twice. Nevertheless, by the age of about four, most children know how language is structured, and the mistakes they make often actually highlight what they have learned. The small boy who says, "I maked a snowball," knows how to form the past tense from the present, though he has been fooled by an exception to the rule.

Partly because all normal children do seem to share a remarkable ability to grasp syntax, American linguist Noam Chomsky believes that a blueprint for language must be coded in human genes, that the structure of language must be genetically determined just as if it were an organ of the body, like an arm or a leg. He points out that all languages are surprisingly similar in syntax, and that children everywhere, as they learn to talk, seem to go through the same steps at approximately the same speed. There is, however, a rival theory. Many behaviorists believe that language is entirely learned, just as swimming is, or any other skill. An infant babbles randomly at first and eventually produces sounds that approximate words ("mama," perhaps, or "papa"). The parents are delighted, and their pleasure serves as a reward, making it likely that the baby will utter the same sounds again.

For anyone who believes that language is a uniquely human ability and that its structure is somehow built into the human brain before birth, the prospect of a talking chimp has to be somewhat unsettling. And, in fact, when the Gardners began to publish news of Washoe's progress, psycholinguists of the Chomsky school were quick to insist that, while Washoe had obviously learned naming, she had not really demonstrated syntax. She was, they said, simply producing strings of signs with little evidence of meaningful structure; in particular, there didn't seem to be any coherent order to her strings.

Word order is very important in English, as I have said, and back in 1970, many psycholinguists believed that, together with voice inflection, word order was universally one of the earliest grammatical devices to show up in the speech of small children. However, new evidence has thrown the whole matter into confusion, because it seems that children learning languages other than

English do not universally follow any particular word order. Furthermore, sign order in ASL is rather different from the ordering of English. It was obviously premature, then, to write off Washoe's accomplishments just because she had not matched the earliest grammatical achievements of English-speaking children.

If the existence of a signing chimp was unsettling for linguists, it was also startling news for animal behaviorists. As I mentioned earlier, before Washoe, it was the general belief that chimpanzees, like all other animals, are only able to convey feelings and intentions. Without language, they are supposedly tied to the present, to the immediate situation, unable to communicate about either past or future. Yet Washoe could sign about the future ("Time eat?" she would ask) and she could talk about things not present, signing "dog," for example, when a dog barked out of sight in the distance.

Are we to believe she was the first ape capable of such mental exploits? The Gardners don't think so. They have written: "As far as we know, there is not a shred of evidence that would support the conclusion that communication among apes is limited to the expression of emotional states." They feel reasonably sure that chimps in the wild must share some form of gestural communication, because it was just too easy to teach sign language to Washoe.

In October, 1970, a little over four years after Washoe's arrival in Reno, the Gardners ended the project. She had acquired 160 signs and had startled the scientific world, but now there were new problems. There was a school across the street from the Gardners' house, and they worried that one day the children might slip into the yard, tease Washoe, and get hurt. There were also rumors of a shopping center to be built soon nearby. In addition, three of the graduate students who had worked with Washoe were getting their degrees and leaving Reno, so the Gardners would have to train new assistants. The two psychologists had learned a great deal from their work with Washoe and they wanted to begin again with new chimps, making the best use of their experience.

And so Washoe was given to one of the graduate students, Roger Fouts, and she traveled with him to the Institute for Primate Studies outside Norman, Oklahoma. There they were both to begin a new life. For Fouts, this must have been an exciting prospect. Since the institute breeds chimps, he would be able to teach signing to a whole cadre of apes; he hoped that eventually the animals would begin to sign to one another.

For Washoe, however, the prognosis was mixed. She would be leaving behind her trailer home and constant human companionship, and she was to be integrated into a colony of chimps. It seems likely that she thought of herself as human, for she had had virtually no contact with her own kind since she was captured as a baby in Africa. It could not be an easy transition.

3.

Lucy and Ally and Bruno and Booee

"Zira!"

I stop and take her in my arms. She is as upset as I am. I see a tear coursing down her muzzle while we stand locked in a tight embrace. Ah, what matter this horrid material exterior! It is her soul that communes with mine. I shut my eyes so as not to see her grotesque face, made uglier still by emotion. I feel her shapeless body tremble against mine. I force myself to rub my cheek against hers. We are about to kiss like lovers when she gives an instinctive start and thrusts me away violently.

While I stand there speechless, not knowing what attitude to adopt, she hides her head in her long hairy paws and this hideous she-ape bursts into tears and announces in despair:

"Oh, darling, it's impossible. It's a shame, but I can't, I can't. You are really too unattractive!"

—PLANET OF THE APES,
by Pierre Boulle

Washoe's introduction to the chimps in Oklahoma was not auspicious. She labeled them "bugs"—and Washoe had little use for bugs.

At first, she signed to humans and chimps alike, although of course none of the chimps knew how to sign back. "Tickle," she would implore the adult chimps, and when they ignored her, she would chase them until they gave in and provided tickling. Once when she was competing with another animal for a handout of fruit, she tried to distract him: indicating a nearby water faucet, she signed, "Go drink." With younger chimps, she most often signed, "Come hug." One day she was grooming a youngster named Bruno when two chimps nearby began to give hoots of alarm and walked rapidly away, probably because they had seen a snake. Washoe started to follow them, but Bruno stayed where he was. She signed, "Come hug, come hug," evidently intending to carry him away from the danger, but Bruno failed to get the message. Washoe finally had to come back and literally drag him away.

The Institute for Primate Studies is a breeding colony owned and run by Dr. William B. Lemmon. It appealed to Roger Fouts as a place to continue his research, for a number of reasons. It is one of the few chimp colonies in the United States where the animals are studied for behavior but aren't used for medical experiments, and it is also one of the few places where these highly sociable apes are housed in groups rather than in individual cages. There are miles of woods on the property, where the more well-behaved apes can be taken for walks, and there is a small island where they sometimes run free. In addition, the institute had placed several infant chimpanzees in human foster homes in the area. Thoroughly accustomed to being around people, they were ideal candidates for lessons in signing.

One of Fouts's first projects after he arrived in Norman was to give ASL lessons to several animals, in an attempt to prove that Washoe wasn't the only chimp who could learn to sign. As he got his study underway, it must have seemed as if a blitz had hit the institute, since he enlisted the help of four young chimps and more than thirty student volunteers. First he taught the students ten ASL signs; then the students taught the chimps.

The results proved that Washoe was not unique. All the animals learned at least some of the signs, and the best of them, when tested, got 90 percent right.

Afterward, Fouts continued to work with two of the chimp youngsters, Bruno and Booee, teaching them identical thirty-six-sign vocabularies in the hope that they would begin to sign to one another. For the most part, they seemed to prefer normal chimp communication, so Fouts had to devise situations that would encourage signing. He tried giving fruit or a drink to one of them and nothing to the other. In these circumstances, however, communication tended to be strictly one-way: the deprived chimp would sign, "Gimme drink" or "Gimme fruit," and the chimp in possession would simply run away. Social separation worked better. When one of the pair had misbehaved and was being scolded and made to sit in a chair, he would often become agitated and sign to his friend, "Come," "Come hug," or perhaps, "Come hurry." If the friend was ordered to stay away, then he, too, would begin to sign.

Fouts also tried requiring them to sign "tickle" before they could begin a tickle-chase game, and soon they began to sign spontaneously in this situation. One day the following conversation was recorded:

BOOEE: Tickle Booee.

BRUNO *(busy eating raisins from the experimenter's hand):* Booee me food.

Fouts also arranged to teach signing to three chimps who were being raised in human homes. The youngest of them, Salomé, was just an infant when she began to have ASL lessons twice a week. By the time she was four months old, she had acquired her first sign. Human children generally don't utter their first words until they are about a year old, though children in deaf families often begin to sign at about five months of age.

All the home-raised chimps Fouts worked with were bilingual: they understood a great deal of spoken English as well as sign language. Chimps with a substantial passive vocabulary had been reported before—Viki, for example, at the age of six understood about a hundred English words. However, Fouts was able to demonstrate that an ape can actually translate from one language to another.

The chimp he chose for the translation experiment was a youngster named Ally, who had been adopted at the age of two months by Sherri Roush, a social worker. Fouts began teaching Ally when he was just over a year old, and by the time Ally was three, he knew more than seventy signs. For the experiment, the psychologist selected a number of common objects—a spoon, a nut, a leaf, for example—that Ally didn't yet have signs for. However, he did understand the English name for each of these, since he would fetch them if asked to in English. Fouts then taught him the signs, using the spoken words alone—not the objects. "Spoon," he said, as he molded Ally's hands in the sign for "spoon." Later another experimenter tested Ally by actually showing him a spoon and asking him in sign language, "What that?" And Ally signed "Spoon," right on cue. This is reminiscent of the way humans learn a foreign language, since when a student of French grasps the fact that *la plume de ma tante* means "my aunt's pen," there is often neither a pen nor an aunt in the room.

Lucy was the third of the home-raised chimps Fouts worked with. Born in January, 1966, and adopted when she was two days old by Maurice and Jane Temerlin, Lucy did not begin her ASL lessons until she was almost five years old; nevertheless, she was one of the brightest pupils Fouts had, able to learn a sign sometimes in a minute or less. In the beginning, he worked with her for one or two hours a day, five days a week. Within two years, she knew more than eighty signs.

Before Roger Fouts came into her life, Lucy, like Ally, already understood quite a lot of spoken English. In his fascinating book *Lucy: Growing Up Human*, Temerlin recalls that if he said to her, "Get a cup and I'll fix you some tea," she would go and get him both a cup and a tea bag. She could also communicate her own needs quite well, using chimp gestures, facial expressions, and vocalizations, all of which her human family understood. When she learned to sign, then, she did not suddenly acquire the ability to communicate; she simply broadened what she could do.

Fouts chose Lucy for an experiment designed to get at the way chimpanzees conceptualize—the way they see the world. He presented her with twenty-four different fruits and vegetables, which she could handle and even eat if she wanted to, and asked her to name them. At the time, she had only five signs in her vocabulary that were related to food: "food," "fruit," "banana," "drink," and "candy."

More often than not, Lucy referred to the vegetables as "food" and to the fruit as "fruit." However, she was sometimes more specific. She called lemons and oranges "smell fruits," and a piece of watermelon was a "candy drink" or on another occasion a "drink fruit." The first few times she was presented with a radish she signed "food" or "fruit" but then she bit into it, and after that it became "cry hurt food."

In a brief digression from his research with chimps, Fouts gave signing lessons to an infant orangutan and to a monkey. Tommy, the orang, did learn a few signs and began to string them together, but Fouts found that teaching him was not as easy as working with chimps, since an orang is by nature a solitary and not particularly sociable animal. The experiment with the monkey, a stub-tailed macaque, was not a success. Fouts finally succeeded in getting the animal to put its hands outside the cage, but then it wouldn't look at him, since to a macaque, direct eye contact constitutes a threat.

The chimps in Norman are learning to sign under very different conditions from those the Gardners provided for Washoe. She was immersed in signing, and her life-style was made as much like that of a human child as possible, since the Gardners wanted to study the development of language. The research Roger Fouts has been doing, on the other hand, generally breaks down into a series of separate experiments, and his chimps get only a few hours a week of signing lessons. Language is not Fouts's first love; he's interested in chimps themselves and their mental abilities, and, as he put it, "you can get at that piece-meal."

The Gardners don't encourage visits by journalists. Though I had the chance to talk to them once after a lecture they gave, I wasn't able to go to Reno to see their new chimps. I was delighted, then, when Fouts agreed to let me come to Norman. At last I would see the signing chimpanzees face to face rather than simply reading about them.

So it was that on a blisteringly hot Sunday afternoon in July, 1975, I checked into a motel in Norman, Oklahoma. The following morning Fouts came by to pick me up. He's a medium-height man with a blond mustache and longish hair, very relaxed and easy to like, and as we drove out to the institute in his car, he gave me a progress report on his chimps. I had read so much about them by that time that, as he talked, I felt like someone getting news of old friends.

Washoe, he told me, had finally made the adjustment to living in the colony. She was ten now and sexually mature, so she could conceive at any time. Although Roger is as eager as anyone to find out whether she will teach signing to her baby, he said he was hoping she wouldn't become pregnant for a while. Even for a chimp, ten is young to be a mother.

Of the three home-raised apes, only Lucy was still with her family, and Fouts was continuing to work with her. Salomé was gone. When her human father died, her mother could no longer keep her, so she was given back to the institute. Despite all efforts to help her, she went into a deep depression and eventually died. Baby chimps are enormously dependent on their mothers, human or chimpanzee, and the institute had lost four infants under similar circumstances.

For a woman or a man with a strong parental instinct, playing "mother" to a chimp must be an incredible experience. Chimpanzee infants want to be held at all times—in fact, they scream if they're put down. In the wild, they cling constantly to their mothers, hardly ever letting go. However, if they are more demanding than human infants, they are also even more responsive and alert. They're no trouble to carry around—they hang on by themselves—and with their bright eyes and wise little faces, they're irresistible. As Fouts put it, they have "hold me" written all over them.

Ally's mother got married, and so he, too, was returned to the institute. At first, he was very depressed. He picked himself bald in places and one of his hands hung limp; he couldn't or wouldn't use it. But he recovered. (Washoe, of course, was older when she made the move from Reno, and in any case, Roger came with her.) When Ally and Bruno were introduced, Ally did a lot of signing, mostly about food and play. ASL is Ally's first language—as it is Washoe's—and he prefers to use it, rather than chimp communication, both with humans and with chimps.

The most exciting news Fouts had was that some of the institute's nonsigning chimps had begun to learn signs from the signers. Pan, who is the father of many of the younger chimps in the colony, now has a few signs, and a three-year-old named Kiko learned "food" and "drink" from Bruno and Booee and was using them correctly when, unfortunately, he died. One day Roger heard that another chimp, Sebastian had been signing "fruit," "drink," and "food," and so he visited him, bringing along all three items, and tried to get him to sign; there was no response. Apparently Sebastian simply saw good things happening to

Washoe when she made certain gestures, but he made no real connection between the gestures and any particular objects.

Manny has also been learning signs from Washoe. Since she reached maturity, Washoe has shown a distinct sexual preference for Manny over other males. When he's in the next cage, she sometimes signs, "Come hug," to get him to come over to the bars that separate them. Manny soon figured out what she meant by that and now signs "Come hug" when he wants Washoe to come closer.

The chimps at Norman have invented a number of signs for themselves. For example, once when Washoe and Roger were out in a rowboat near the chimp island, Roger indicated a pair of swans and asked her "What that?" He had always referred to them before with the sign for "duck," so her answer astounded him: she called them "water bird."

But Washoe's most startling invention was a form of insult. One day Roger took her into a barn at the institute where several kinds of monkeys live in cages; he planned to teach her the sign for "monkey." Before he could begin the training, however, Washoe started threatening a macaque, who threatened back. Roger stopped the argument, showing her a siamang (actually a small ape) in another cage, and molded her hands in the "monkey" sign. After he had done this several times, he indicated a cage with squirrel monkeys in it, and she quickly made the connection and began to refer to them, too, as "monkey." But when he turned at last to the combative macaque and asked, "What that?" she signed, "Dirty monkey." Washoe had previously used the sign "dirty" only to refer to feces or to something soiled, but from that day on, she used it as an insult, applied to people who earned her displeasure. "Dirty Roger," she signed when she asked for fruit and he didn't supply any. Lucy had used the same sign in the same way, signing "Dirty cat" for a strange cat. Her leash, which she dislikes, is sometimes a "dirty leash." Lucy has also invented her own sign for "leash": she holds her index finger, bent like a hook, against her neck.

As Roger talked—his affection for his chimpanzee pupils is quite obvious—we were driving along narrow roads that shimmered with mid-morning heat, heading out into the country. At last, we turned into a private drive, and the institute lay ahead of us. It looked at first like any small farm, with a rambling main house and some outbuildings, including a barn. Just then there was a roar that reminded me of lions in the zoo: it was Washoe, greeting Roger. When we got out of the car, there she was in the cage that occupied the whole roof of the one-story building Roger called the lab. She signed at us and Roger translated for me: she was asking for fruit.

Straight ahead of us was a small lake with several islands in it. On one of them, four gibbons could just be seen through the thick foliage of the trees. The next island over was quite fabulous, a tangle of lush green shrubbery with a pink concrete rondeval—a round hut reminiscent of Africa—in the center. Two small chimps came hurrying over to the edge of the island. One of them—it was Ally—signed hopefully, "Come out."

The lab had a complex of cages inside. They were arranged in modules, like the rooms of a railroad flat, with indoor space connected to outdoor space by hatchways, so that if there was a dispute, the disputants could get away from one another.

Pan was in the first cage, and as we paused in the doorway, he doused me with a mouthful of spit—something captive chimps are prone to do. Bruno, Booee, Manny, and another male named Pancho were in the next cage, and in another one, farther down the room, were several female chimps and their babies. The infants, with their pale little faces, really *were* irresistible, but I was somewhat taken aback to see that the mother chimps had flat but very human-looking breasts. Absurdly, I had somehow expected them to feed their babies from teats on their abdomens, as other mammals do. In a nearby cage was a chimp who was disturbingly human in a different way: a retired circus performer, she walked erect on two legs, looking for all the world like a hairy, waddling old woman.

After the tour of the lab, we settled down in chairs on the lawn for a while. In one of the outdoor cages, a juvenile chimp about the size of a human four-year-old was playing by himself. Putting one hand over his eyes as a blindfold, he began to twirl slowly around and around until he was staggeringly, deliciously dizzy.

I asked Roger about the dangerousness of chimps, because I had heard that Washoe was now too dangerous to work with.

"It's standard lab lore that chimps six to eight years old and older are unsafe to be around," he told me. "And it's true that caged animals *are* unsafe—in San Quentin or the local zoo. It's also true that you can't intimidate an older chimp as easily as you can a young one—just as you can't treat a fifteen-year-old human like a two-year-old. As a chimp grows older, your methods have to change."

Washoe now weighs around 100 pounds and is as strong as a 500-pound human, but Roger is still working with her. However, these days he doesn't just order her around; instead "we sit down and discuss things." He explained that chimps, like children, are always testing the limits. For human safety, there have to be limits, but especially with older animals Roger avoids a battle of wills, and before an argument can really get rough, he will generally change the subject. For example Booee went up a tree one day when Roger had him out on a lead. It was time to go back to the chimp island, so Roger tugged on the lead—and Booee reached down and started to lift him right up into the tree. "I did what any good psychologist would do," Roger said. "I signed, 'You're forgiven, come hug,' and he came down."

Fouts stressed that to handle chimps, you also have to know the individual animal. Washoe, for example, is very food oriented, which can be useful in teaching her, but if Roger tried to use the same methods on Lucy he would get very fat, since she would rather watch him eat than do the eating herself. Washoe is something of a tomboy; Lucy is more refined; Ally is hyperactive and rather sensitive—he gets so involved in the language training that he becomes upset when he doesn't get his answers right.

Bruno and Booee are also quite different. Bruno can be sneaky, but he takes a scolding quietly if he knows he has done something wrong (such as stealing Roger's pipe and retreating up a tree). He can be reasoned with and is open to compromise. Booee, on the other hand, is very impulsive and, if scolded, is apt to explode. However, he's also paranoid, and Roger takes advantage of this: when he wants to end an argument with Booee, he changes the subject by giving a chimpanzee alarm bark. Booee immediately runs to him for protection, forgetting the dispute they were having.

I also asked Roger about Washoe's progress and the present size of her vocabulary. She still had, he told me, only about 160 signs, although he was planning to teach her 20 more soon as part of an experiment. Someday he would really like to let her vocabulary take off, but for the time being, because of experiments he plans, he needs to control exactly which signs she does and doesn't know. Lucy and Ally both have about 100 signs each. Bruno and Booee were supposed to be limited to 38 signs each, but they're up to 44 now because they've been learning from Washoe and from watching humans sign to other chimps.

"You just can't stop chimps from learning," Roger said. "All I do is teach vocabulary, individual signs; I can't take much credit for that. Give a chimp four signs and he starts combining them on his own."

He explained that vocabulary size isn't all that interesting, anyway. A chimp could know sixty thousand signs, yet they'd be just labels without syntax. "It's how they combine signs to express themselves that will show the mentality of the chimp—will give us an idea of what goes on behind those brown eyes." He paused thoughtfully. "We can learn a lot from the similarities *and* the differences between chimps and humans. For example, I'd like to find a strictly chimp characteristic to Lucy's language. After all, a chimp is a chimp, and much too delightful a species to become us."

I asked Fouts whether any of the signing chimps ever use words to express or describe their emotions, and he mentioned that Lucy knows the signs for "cry," "sorry," "happy," and "smile," and uses them appropriately. For example, one day recently when she was eager to go outside, Roger teased her: he took her leash, hooked it onto his own collar, went to the door, and signed, "Out, out hurry"—which is what Lucy usually does. She played along and took the part of Roger, signing, "In." He retreated to the couch, sat and rocked and signed, "Cry me." But Lucy had evidently figured out that she wouldn't enjoy the outcome of this little game, so she signed back, "Cry Lucy." He said, "No, cry Roger," and she one-upped him with, "Hurt Lucy."

There is, of course, no way to prove that when a chimp signs "Cry me," she is feeling the same thing a human might be feeling who used similar words. One can only empathize; but then, even among humans, we have to take the other person's word for it that she feels what she says she is feeling.

We talked a bit about anthropomorphism then, about ascribing human traits and feelings to animals. This has long been considered a cardinal sin by those who study animals. Yet Roger finds that he is constantly responding to cues from his chimps that are hard to analyze or describe. He does empathize, and he says he has seen all the human emotions in chimps—love, anger, skepticism, guilt, and much more. With a species that's so much like us, empathy and anthropomorphism may be a good thing, since if you rule them out and study chimps using the same carefully objective techniques that are appropriate for studying simpler creatures, such as birds or fish, you reduce drastically what you can learn about the animals.

When Roger converses in signs with a chimp, he reads into the message the ape's facial expressions, postures, and eye contact or lack of it. In fact, a friend says that what Fouts has created in his work with chimps is first-generation pidgin—a blend of Chimpanzeese and human communication—which will eventually develop into a kind of creole. Roger would like to be able someday to consciously integrate naturally occurring chimp signals with signing, to do deliberately what he has been doing unconsciously and unintentionally all along. He promised to introduce me later to Sue Savage, a graduate student who was studying normal chimp communication at the institute.

I asked Roger whether any of the chimps had ever lied to him. It's an interesting question, for supposedly humans are the only animals capable of lying, because we are (or were) the only animals capable of language. It turned out that Lucy *has* been known to lie. One day when Roger came to give her her language lesson, there was a bowel movement in the middle of the floor. Obviously it was Lucy's, since chimps are notoriously difficult to toilet train and Lucy was not always reliable. Deciding to take a stern tone with her, Fouts signed, "What that?" Lucy looked away, pretending not to see the question. The conversation went on from there:

ROGER: You do know. What that?
LUCY: Dirty, dirty.
ROGER: Whose dirty, dirty?
LUCY: Sue's.
ROGER: It's not Sue's. Whose is it?
LUCY: Roger's!
ROGER: No! It's not Roger's. Whose is it?
LUCY: Lucy dirty, dirty. Sorry Lucy.

The ability to lie, however, isn't something chimps have acquired as a kind of dubious fringe benefit as they learned language. It's also possible to lie without words, and Fouts recalled an incident that occurred one day when Bruno and Booee were on the chimp island. The inside of the rondeval was being hosed down, and Bruno began playing with the hose. However, Booee, who is bigger, soon took it from him. At that point, Bruno went to the door and gave an alarm bark, and Booee rushed outside to see what was worrying his friend. Bruno immediately seized the hose. At first, the humans present assumed, along with Booee, that Bruno actually *had* seen something alarming. However, when Booee returned and reclaimed the hose, and Bruno gave the same bark again, and the whole sequence was repeated twice more, they realized that he was lying. He had learned Roger's technique for distracting Booee.

I was about to ask Roger more questions when Sue Savage appeared, and Roger said he had to get back to work. Sue is small, sandy-haired, fresh-faced. She hardly looked a match for an adult male chimp who was, Roger had assured me, the equivalent of a strong seven-hundred pound man; yet she was there to walk Pancho, a nonsigning adult chimp. Sue's primary interest is in the way chimps normally communicate.

She brought Pancho out of his cage on a very long leather lead. As they passed me, he decided to get up on the lawn chair beside mine. He sat there for a while, staring straight ahead, while I watched him out of the corner of my eye, trying not to look as nervous as I felt. Eventually he hopped down and ambled off to the barn for a snack of Monkey Chow. Then he and Sue and I, together with two young women who were graduate students working with Sue, settled on the bank above the lake, looking across at the chimp island, while Sue described the way Pancho makes his needs known. If he wants more food or drink, she said, he will sign like this: "You (pointing to her) go (sweeping gesture) get more (holding up his empty cup or a grape stem)." Sometimes he simply shoves her away and makes food grunts: "Uh-uh-uh." Either way, his meaning is clear.

As we sat there talking—and I stopped worrying about Pancho, who was sitting quietly on the grass nearby—the sky abruptly clouded over, and within seconds, it began to pour. We all headed for the barn at a run. Sue stopped to roll up her car windows, and Pancho instantly climbed into the car and refused to come out, so she got in with him.

Inside the barn, the rest of us stood around waiting for the rain to let up, while back in Sue's car, Pancho became increasingly restless. Finally she opened the door for him and they made for the barn. Pancho, however, became very excited, and his dash for shelter turned into what may have been a rain display, for he came on like an express train, his hair on end, and as he reached the barn door, he leaped into the air. Rising fully four feet off the ground, he made several great bounds and hurtled past us toward the other end of the barn. Suddenly he no longer looked like an unusually hairy human, the illusion shattered by his incredible strength and uncontrollability.

The famous ethologist Jane Goodall, in her studies of chimps in the wild, has described behavior similar to Pancho's. She watched more than once during a rainstorm while adult male chimps charged down a hillside over and over again, hurling tree branches ahead of them in a spectacular display.

Was something instinctive and irresistible in Pancho triggered by the rain? I began to feel that the fascination of working with chimps must come partly from their humanness and partly from the fact that, of course, they really *are* a different species. A relationship with a chimp demands all the same things a relationship with another human does, only more so; at any given moment, it takes all one's energy, attention, and emotions.

When the rain finally stopped, we set off for a walk through the woods with Pancho. For me, it was a fantastic, almost unreal experience. As we trudged along, with rain dripping on us from the trees and clouds still scudding overhead, it was easy to imagine that I was Jane Goodall, following chimps at the Gombe Stream Reserve. Up ahead, Pancho knuckled along through the red Oklahoma mud on the heels of Ricky, one of the graduate students. At one point, they sat down on a log facing each other—seated, they were the same height—and Pancho groomed Ricky, gently removing sleepy seeds from her eyes, running his hand over her hair, motioning her to turn around so that he could groom her back, making teeth-clicking noises throughout. After a while, it was her turn to groom him. Then we moved on again. We tramped through the mud for about an hour before it was time to go back to the institute, where Pancho reluctantly reentered his cage. Cages, Roger Fouts had said, are hard on chimps and hard on people.

The following day, I had a last interview with Fouts and we

talked about his plans for future research. Eventually Washoe or Lucy will have a baby, and then the question of whether she will teach signing to her offspring will begin to be answered. It seems not at all unlikely. For a few months, Lucy had a pet cat that she carried everywhere with her; when she first got him, she tried to teach him to sign. She would put him down in front of her, show him an object, and sign, "What that?" Then she would sign the answer to her own question, imitating the language lessons she was having with Fouts. When, after several days of "lessons," the cat still did not sign back, Lucy abandoned the project.

Though chimps in the wild apparently don't instruct their young, Sue has seen mother chimps at the institute demonstrate for their babies how to get water from a faucet; and she has seen a mother wrap an infant's small fingers around the bars of the cage, apparently teaching it how to cling there.

In a study related to teaching, Roger hopes to work with baby chimps temporarily separated from their mothers. He plans to use a combination of chimp communication and signing to establish a relationship with the infants, and then he will teach them some signs and a few techniques of tool use. After that, he will return them to their mothers and watch for the spread of both signing and tool use. Among primates, there is precedent for the infant as teacher. At the age of eighteen months, Imo, a Japanese macaque who was part of a captive troop, learned to wash potatoes in a brook before eating them; soon the other monkeys were doing it, too. At the age of four, she solved another problem, one that was created when experimenters scattered grains of wheat on a sandy beach. She took a handful of wheat and sand and threw it into the water: the wheat floated and the sand did not. Again, the rest of the troop learned the technique from her.

Further into the future, Fouts would like to study predation. He and Dr. Lemmon, the institute's owner-director, hope to turn loose a posse of signing chimps in a fifty-acre enclosure, together with a small flock of sheep, to see whether the chimps will hunt the sheep. In the wild, chimpanzees are for the most part vegetarian, but they do occasionally kill and eat other animals such as baby baboons. Since some scientists insist that humans developed language when they began to hunt, or else that it was language that made cooperative hunting possible, this might be a revealing experiment.

I asked Fouts where he thought it would all end. How much language can chimps learn, and what will happen if someday there are whole populations of signing chimps? He said he simply has no idea. "Every time anyone says chimps can go this far and no farther, they prove otherwise. We don't know what they're capable of."

Someday, he would like to see chimpanzees respected as members of a different culture. However, given the human inclination to exploit rather than to respect, he is not optimistic about this.

In August of 1976, a year after my visit to Oklahoma, Washoe had a baby. Unfortunately it was born with a severe congenital heart defect and only lived about four hours; but before it died, she signed to it.

"All the maternal behavior was there," Roger Fouts told me over the telephone. "She was very, very good with it. When it stopped reacting, she would lay it on the bed and then sign to it, either "hug" or "baby." Then she would pick it up and hold it and try to get it to respond. It was very sad, but she's all right now. Hopefully she will become pregnant again."

4.

The Younger Generation of Signing Apes

"And you carry out these experiments on men!"

"Of course. Man's brain, like the rest of his anatomy, is the one that bears the closest resemblance to ours. It's a lucky chance that nature has put at our disposal an animal on whom we can study our own bodies . . ."

—PLANET OF THE APES,
by Pierre Boulle

In the ten years or so since the Gardners first began to work with Washoe, psychologists have learned a great deal about apes and what they can do with signing. However, we still don't know what the outside limits are, because none of the animals has yet reached maturity—in 1977, Washoe was only twelve. Chimps are said to mature intellectually at the age of about sixteen and, in captivity, sometimes live well into their forties.

While the members of the first generation of signing apes are still learning and expanding on what they can do, some of their teachers have gone on to a second generation of language experiments, and new researchers have entered the field, as well. Much of the work is haunted by two basic problems. It's difficult to find people who are fluent in sign language who are also comfortable handling apes—especially as the animals grow older and stronger; and then there is the struggle to secure adequate funding, since chimp language experiments are expensive to sustain. As I talked to the psychologists involved in some of the older projects, the work sounded fantastically exciting and rewarding. It wasn't until I began to interview people who were in the beginning stages, teaching signing to very young chimps, that I realized what extraordinary patience the research demands. Epic problems sometimes come with the territory, and there can be heartbreak as well.

After Washoe's departure, the Gardners began again with two infant chimps. They also changed their training procedures in significant ways: they started exposing the animals to ASL within just a day or two of their birth; and they placed each with a family of native signers, either a deaf family or a couple whose parents had been deaf. Washoe, by contrast, was already eleven months old when she arrived in Reno, and her family—the Gardners—were beginners in sign language.

Whether it was the early start, the exposure to fluent signers, or both, the new procedures quickly paid off. Moja, who was born in late 1972, was just three months old when her first four signs appeared. Pili, born about a year later (the names are Swahili for "one" and "two"), formed his first sign when he was fourteen weeks old, and within a week after that, he too had a four-word vocabulary. By the time they were six months old, Moja and Pili each knew more than a dozen signs; after six months of training, Washoe knew just two.

Unfortunately, Pili died, which must have been a cruel blow to the humans who worked with him. Subsequently, the Gardners have acquired three more chimps, the youngest just eight months old, and the project is continuing.

Back in 1975, when I was in the early stages of researching this book, I was delighted to discover that there was actually a signing chimp right in Manhattan, just a subway ride away. The chimp, Nim, was adopted by a New York family when he was just a few days old, and spent the first couple of years of his life in a West Side town house where he had regular signing lessons. Later, he was moved to suburban Riverdale, New York; from there he commuted by car to his ASL classes at Columbia University three times a week. Nim was born at the institute in Oklahoma, and a glance at his family tree suggests the era of signing chimps is well under way. He's the son of Pan and Carolyn, brother of Ally and half-brother of Bruno, and he's the fourth of Carolyn's children to be taught to sign. (As of this writing in late 1977, he's back at the institute in Oklahoma.)

The primary moving force behind the Nim project was Columbia psychology professor Herbert Terrace. I talked to him once in 1975 and again in the spring of 1977, and after the second interview, he invited me to watch one of Nim's signing lessons.

Deep in the bowels of the natural sciences building at Columbia, a suite of small rooms had been set aside for Nim. There was a chimpnasium fitted with climbing equipment, a kind of utility room, a classroom, and an observation room from which we watched. As we entered quietly, I could see through a window of one-way glass that Nim was already in the classroom, sitting in a high chair eating cereal with a spoon. He was wearing a white T-shirt that had a sketch of a chimp's head on it above the words *Bronx Zoo* and he was handling that spoon with considerable aplomb. Professor Terrace had explained to me earlier that wearing clothes gave Nim a whole extra category of things to sign about and was part of his socialization as well.

The classroom was quite small, comfortably carpeted but bare of furniture except for the high chair and a few shelves piled with toys and books. Several posters had been taped up on the walls. One showed a chimp dressed much as Nim was, sitting on a toilet eating a banana. Another showed a cat sitting on a toilet and the legend underneath read, Nothing is impossible. Since Nim had recently been toilet trained, I suspected that the posters were propaganda. However, according to Terrace, they were only there to provide yet another topic of conversation.

As Nim finished his cereal, the trainer working with him at the moment, a tall young man named Richard Sanders, signed to him: "Thirsty" and then "cup." With a little encouragement, the chimp repeated both signs and was given a mug of juice. He drank it down in no time, signed, "Dirty, finished," and, several times, "Dick hug." Then he stood up in the high chair and held out his arms to be picked up. Like a human preschooler, at three and a half, Nim was a good, solid armful.

Out in the utility room, he hung from Dick's neck while Dick washed the breakfast dishes; occasionally he lent a hand. In Riverdale, Nim shared a large house and thirteen acres of grounds with four humans who took every opportunity to sign to him. Part of the house was reserved for the humans; the rest was Nim's, and the two areas overlapped in the kitchen, where Nim sometimes helped with the chores. He loved to stir cake batter and wash dishes, whether they actually needed washing or not. A few months earlier Nim had to make a difficult adjustment when his original signing companions were replaced by new people. Professor Terrace was having funding problems, and the money he needed didn't come through in time for him to hang onto his original staff. Nim was upset at the change and actually tried to run away from home—slipping out the front door and dashing off down the street. However, he eventually adjusted, just as he had adjusted to the move from Manhattan to Riverdale and the separation from his original human family. Terrace decided on that move because the new arrangement would give Nim more exposure to signing and more room to grow in.

Back in the classroom, Dick took a child's book from a high shelf and sat down on the floor while Nim cuddled up to him. The book was filled with large, bright-colored photographs of simple things—a spoon, a cup, gloves,—and Dick set to work to get Nim to sign the names of some of the items. However, the chimp was restless. Several times he scrunched down, putting his nose right up against the photograph in the book; or he would squirm, looking anywhere except at his teacher, until Dick took his chin and turned his head gently so that he had to follow the signs Dick was making. Once Nim signed "sleep." Terrace explained that when he wanted to get out of the classroom or out of any situation, he would often sign "sleep" and then lie down, looking up out of the corner of his eye to see if his teacher was watching.

"One thing you have to understand about chimps," Professor Terrace told me, "is that they don't sign as spontaneously as a child talks. You have to remind them to do it by asking, 'What's that?' or signing 'Talk to me.' It's not that they're not smart enough, but their motivation is different. They're more creatures of impulse, of motor action. We have to persuade Nim that it's worth his while to sign."

From time to time, as Dick worked with Nim, he would stop and murmur into a small tape recorder, reporting on what had just taken place. And from time to time as Nim produced a good, clear sign, Terrace would photograph him with a camera hidden in the wall of the room. In this way, and with videotape, Terrace is compiling a record of Nim's progress.

As we left the observation room to go back upstairs to Terrace's office, I mentioned that Nim seemed to need a lot of body contact. He was enormously affectionate and was snuggled up to—or at least kept a hand on—Dick most of the time. Professor Terrace said that that was good, because it meant he was more dependent. From the beginning, Terrace went to great lengths to create an extremely socialized chimp, something that seemed particularly important because chimps often become unmanageable during adolescence and male chimps are much stronger than females. Terrace was very encouraged by the fact that Nim often signed instead of acting on his impulses: he would sign "bite" or "angry" instead of actually biting. Learning to substitute words for physical aggression is, of course, an important step in the socialization of young humans as well.

Summing up Nim's progress, Terrace told me that he had a vocabulary of ninety-two to ninety-five signs and that, to date, he had strung them together in 4700 different combinations of from two to five signs. The number of combinations was increasing much faster than his vocabulary was. This was significant because Terrace wanted to prove that Nim was not simply producing rote sequences, something even a pigeon can learn to do. Since it would be virtually impossible for him to have learned 4700 different sequences, and since many of them were his own invention anyway, the point seems to have been made. However, it was also necessary to demonstrate that Nim's combinations weren't simply *word salads*, a matter of stringing together a handful of signs that were all appropriate to the context. This, of course, is the syntax question again. Terrace had been looking for patterns in Nim's combinations, for some indication of meaning expressed through structure. So far he had found that Nim always put the "more" sign before the word it modified ("More tickle."

"More banana.") and he always put verbs before their objects ("Tickle me." "Hug Nim.") In fact, he had now built that into a patterned three-sign sequence and would say, "Herb tickle me," "Joyce hug Nim," and so on.

"There's no question," Terrace said, "that he has a number of very simple grammatical rules with which he can generate sentences. Otherwise it would be impossible to account for the regularities in what he's doing."

In late 1977 Nim had to be sent back to the institute in Oklahoma, largely because Terrace once again ran into funding problems. Presumably, Nim's ASL lessons will be continued and in addition he'll be able to sign to other chimps such as Washoe; but for a youngster used to constant human companionship and a fair degree of physical freedom, Oklahoma is bound to be something of a comedown.

In the fall of 1976, I managed to visit Koko, the signing gorilla who lives at Stanford, for a second time. (I first saw her in July, 1975, an experience described in Chapter I.) She had grown a lot—Penny Patterson's guess was that she had doubled her weight within the last year—and was as outgoing as ever. As soon as I arrived, she came over to the trailer's wire-mesh fence to push her lips out and blow at me. Remembering that the last time I was there, Penny had explained that she did that because she was curious about what I'd been eating, I blew back and said, "Peanut butter sandwich." I was just telling her how good it was to see her again when she brought her fist up to her mouth in what was obviously the sign for "nut." I was floored. Like a lot of people, I often talk to animals, always with the comfortable conviction that they can't comprehend a word I'm saying. That was what I had been doing with Koko—forgetting that she could understand—when she reminded me that she was different. On some deeper level, it's difficult to absorb the fact that it's now possible to converse with apes.

Penny, who is a tall, slender, fair-skinned blond, was looking more beautiful and more fragile than ever. It was a rough period in her life. Officials of the San Francisco Zoo had begun to say that they wanted Koko back (legally, Koko belonged to the zoo and had been on loan to Penny for a little over three years). To me—and obviously to Penny—the idea of confining Koko to a cage, to be stared at all day by humans, was unthinkable. The zoo would accept a substitute gorilla, but young gorillas are not easily come by. Penny thought she had a candidate but wasn't sure yet whether she would be able to manage the swap.

Meanwhile she had acquired two baby gorillas, and it might be possible, if all else failed, to give one of them to the zoo. However, they were very attached to one another, and if they were separated, they might both die. They were also both sick at that time, one with a cold, the other with a serious fungus infection. At first, Penny was told that she musn't go near them because she might catch the infection and give it to Koko, but in the end, she decided that she'd have to chance it. With only herself and one full-time assistant to look after three gorillas, she really had no choice.

She was also struggling to finish her Ph.D. thesis, a difficult task since she was frequently distracted by the need to care for her charges, to write up grant proposals, and to hold off the zoo. If Koko's future was up in the air, then Penny's was as well. She had, however, established a nonprofit gorilla foundation and was hoping for donations to help support her research.

Koko was learning as fast as ever and was up to 450 signs, Penny said, though that's not official since the very size of her vocabulary has made it difficult to prove that she has mastered a given sign. Penny has been using the Gardners' criterion: Koko must use a sign spontaneously at least once a day for fifteen days before it's officially counted as part of her vocabulary. With an animal who signs as rapidly and fluently as Koko does, those crucial once-a-day utterances are hard to pick up.

What seems most important about the Koko project is that it has proved that gorillas can learn to sign at least as readily and rapidly as chimps can. Koko has invented signs and phrases; she "talks" to herself as she leafs through magazines; and she understands so much spoken English that Penny occasionally resorts to spelling out words when she wants to keep something from her. The parallels between Koko's accomplishments and those of the chimps are quite striking.

Penny and I talked, sitting on the floor with the wire mesh between us, while Koko beat her chest, thumped on the walls, and did a kind of whirling dance in what was obviously a bid for attention. Giving up finally, she plopped down in Penny's lap and, lolling comfortably, eyed me through the mesh. Becoming restless again, she picked up one of her dolls. She has a whole collection of them, including a bright-red *Planet of the Apes* ape and a rather homely black plastic gorilla. Penny said that she sometimes mothers the black gorilla doll, pretending to nurse it (she has seen human mothers nursing their babies) or putting it on her back, as gorilla mothers do when they need to ferry an infant from one place to another. When she thinks no one is watching her, Koko signs to her dolls, telling them they've been good or bad, asking them to "chase-tickle." She seems embarrassed, though, when humans eavesdrop and will often carry the dolls to the very back of her room, out of sight.

Penny had noticed recently that at regular twenty-eight-day intervals, Koko became moody and irritable. This could be the psychological beginning of the estrus cycle. Though she was only five and a half, she might reach puberty as early as six. Like chimps, gorillas can become hard to manage at adolescence and, given their size and strength, the problems are potentially greater. However, for the moment, Penny is simply planning to keep visitors away whenever Koko is having her period.

Playing mother to an ape is an incredibly demanding job, especially for a single parent. Though Penny has an assistant now and volunteers spell her for brief periods, it's still basically a seven-days-a-week, all-day-and-night commitment. And always there is the worry about what will happen to Koko if the zoo insists on taking her back, or if the project's funding runs out.

It was actually Professor Terrace who first mentioned to me that, sooner or later, the chimp language projects may raise some uncomfortable ethical questions.

"We have never been very concerned about the rights of animals, aside from humane considerations," he said. "But supposing a chimp says, 'I don't want to be operated on,' or 'I don't want to be in a cage,' or even, 'I believe in Jesus Christ'? We have often equated language with civilization; suppose now chimps come up with a civilization of their own? I don't know what the answers are."

But there's another ethical question as well: what is to become of the signing chimps in the end? When the lessons are over, when they've shown what they can do, will they wind up as curiosities, consigned to a zoo? Given present methods for funding scientific research, the ape language projects appear to be terribly vulnerable, because they can be so expensive to run. It seems to me that society has some sort of continuing moral obligation to provide a decent and stimulating life for these animals who have been raised to live like humans.

One very positive development, however, is the fact that there are a number of apes in zoos right now whose lives are being immeasurably enriched by ASL lessons. The Portland Zoological Gardens, for example, has six young chimps enrolled in signing classes as part of a project begun in June, 1972. Every day half a dozen human volunteers come to the zoo to take the youngsters out of their cages. In warm weather, signing lessons take place outdoors, and teachers and students also spend some time wandering around the zoo, visiting other animals. When the weather is cold, the time is spent indoors in a playroom instead. Research is not the goal; this is simply an attempt to provide the animals with an antidote to boredom.

Since the Gardners first began to publish reports of Washoe's progress, an argument has raged over whether signing apes can actually be said to have language. Some critics insist that what Washoe and the others have done can be explained as a complex kind of naming and that there is little evidence of syntax in their strings of signs.

The Gardners have demonstrated that Washoe's signing showed some predictable patterns, and Herb Terrace found patterns in Nim's sequences as well. Roger Fouts has pointed out that many of the chimps he works with can distinguish between "Washoe tickle Roger" and "Roger tickle Washoe" (or Lucy, or Ally, or whoever), and that's certainly a demonstration of meaning achieved through word order. He also cites some of Washoe's longer strings that clearly show order and syntax. For example, one day Washoe wanted a drag on Roger's cigarette. She asked for it again and again, frequently rephrasing her request. "Give me smoke," she began, and then, when he held out on her, she went on to "Smoke Washoe," "Hurry give smoke," and "You give me smoke." Finally Roger signed, "Ask politely," and she immediately responded with "Please give me that hot smoke."

Even more significant, perhaps, was the experiment in which

Fouts taught Ally prepositional phrases—"baby in hat," "hat on box," that sort of thing. Ally soon reached the point where, though he might make errors in the signs themselves, he virtually always got the word order right. In this instance, syntax was obviously not as difficult for him as vocabulary was: all he had to do was perceive a relationship and then go on using it. Fouts suggests that syntax might not be as hard to master as linguists would have us think.

Nevertheless, anyone reading a transcription of a conversation with a signing ape will be struck by the awkward, telegraphic quality of the sentences. "Hurry give smoke," Washoe demanded, and Bruno remarked, "Booee me food." This awkwardness comes largely from the fact that the translations are literal ones; Spanish, translated word for word into English, sounds awkward, too, since the syntax is different. Ameslan also has its own syntax; and word order, in particular, operates differently since American Sign Language is a visual, not a spoken, language. In normal conversation, signers use space to some extent to recreate the event they're discussing, and so the word order tends to be the order of the events themselves.

It's only in recent years that linguists have begun to study how signing children acquire language. The results of one major research project were due to be published in 1977 and promise a chance to make detailed comparisons of the linguistic progress of signing chimps and signing children.

But in the meanwhile, the evidence for chimp language is piling up, not just from projects with signing chimps but from experiments that have taken an entirely different approach, also. We know now that sign language isn't the only kind of language chimps can learn, because since 1965, they have also been taught two quite different artificial languages. In one, the animal operates a keyboard that, in effect, types out word-symbols. In the other, the symbols are bits of plastic backed with magnets, and the ape arranges them in sequences on a magnetic board.

The inventors of these artificial languages believe they have several advantages over signing. For one thing, they do away with some of the ambiguities of sign language. ASL can be tricky to interpret for someone (such as the average psychologist) who is not a native signer. It's sometimes difficult to tell where one sign ends and another begins, and it's easy to read an irrelevant gesture as a sign, or to fail to pick up a sign, thinking it's just an extraneous hand movement. An artificial language can also be designed to be more like spoken English, thereby making comparisons easier, and in addition, it gives the psychologist control over the ape's language production. The animal can, after all, use only the words she is handed or those made available through her keyboard. Furthermore, because the experimenter does have control, she or he can demand correct syntax from the animal. The goal, of course, is to prove, once and for all, that syntax is not beyond the capabilities of an ape.

5.

Chimps Who Read and Write: Sarah

The other project was a scheme for entirely abolishing all words whatsoever. This was urged as a great advantage in point of health as well as brevity. For it is plain that every word we speak is in some degree a diminution of our lungs by corrosion, and consequently contributes to the shortening of our lives. An expedient was therefore offered: since words are only names for things, *it would be more convenient for all men to carry about them such things as were necessary to express the particular business they are to discourse on. . . . I have often beheld two of those sages almost sinking under the weight of their packs, like peddlers among us; who, when they met in the streets, would lay down their loads, open their sacks, and hold conversation for an hour together; then put up their implements, help each other to resume their burdens and take up their leave.*

—GULLIVER'S TRAVELS,
by Jonathan Swift

Ape language projects were obviously an idea whose hour had arrived.

At the same time that the Gardners were gearing up for the

Washoe experiment, a California chimp was being enlisted in a language project based on a very different approach. Professor David Premack, then of the University of California in Santa Barbara, set out to teach his animal to write with bits of colored plastic on a magnetized board. Each plastic symbol represented a whole word: "banana," for example, was a pink square; "give" looked like a green bow tie; "pail" was a red zigzag. The symbols were backed with metal so that they would cling to the board, and Sarah, the chimp, was taught to arrange them in meaningful sequences.

Professor Premack felt that his system had several advantages over signed language. When he asked Sarah a question by putting a sentence on the board, the question remained in place. If she couldn't answer it, he knew it wasn't because she had forgotten what the question was. And because she didn't have to learn how to form signs, like Washoe, or how to form words, as human children do—because all she had to do was pick up a piece of plastic and put it on a board—she could get involved in the training immediately. There was no need for a prolonged period during which human teachers struggled to capture and hold her attention while they demonstrated what she was to do.

Premack is not primarily interested in chimpanzees. "Chimps are fascinating," he told me during an interview in Santa Barbara, "but my main interest is people." What intrigues him most is the nature of intelligence; and language, of course, reflects intelligence.

Sarah, who was born in Africa, was about six years old when Professor Premack began her education. Before that, she was, as he put it, "mostly just being raised: being bottle-fed, taken for walks, spoiled." Once her language lessons began, they were fairly intensive. Sessions lasted from half an hour to forty-five minutes, and she had four a day when she was willing—which wasn't always. Unlike Washoe, Sarah lived in restricted conditions in a wire cage on the university campus. She was supplied with toys but had nothing like the amount of stimulation Washoe received. Nevertheless, within about four years, Sarah had learned 130 words and—more important—could use them grammatically in sentences that were fairly complex. Like the signing chimps, she sometimes talked to herself in her new language. When she became bored with her lessons, she would steal the words from the experimenter, go off into a corner of her cage, and use the symbols, laid out on the floor, to ask herself questions and answer them.

What Premack did in developing a training system for Sarah was really quite remarkable. He virtually took language apart, breaking it down into a series of skills that she could acquire by the simplest, easiest steps possible. He began by teaching her the symbols for "banana," "apple," and "orange." Sarah learned, for example, that when she was shown an orange, she was to pick up the symbol for "orange" and put it on the board. Whenever she did this correctly, she was rewarded with a piece of an orange.

As soon as she knew the plastic names for the three fruits, she was taught the names of her trainers. To make this lesson easier, the trainers wore their name-symbols as pendants around their necks. Now to earn her fruit, she had to write "Mary apple" or "Randy apple," picking the correct symbols both for the fruit and for the person who was with her. Then she was given a name pendant of her own to wear. Soon she was writing whole sentences, such as "Mary give apple Sarah." She wrote in a vertical column on her board; this was her preference, and her trainers (they included Premack's wife, Ann) went along with it.

Because it would be much easier to teach Sarah language once she knew what a question was and that she was expected to provide an answer, Premack made this one of her earlier lessons. She learned that whenever she saw the question-mark symbol, she was to take it off the board and replace it with the word that belonged in that space. Presented with the sequence, "? color-of apple," she would write, "Red color-of apple." If, instead, the question read, "Red ? apple," she would fill in "color-of" where the question mark was. If the question mark came at the beginning of the sentence ("?Red color-of apple"), she knew it had to be replaced either by the symbol for *yes* or by the symbol for *no*.

All Sarah's tasks, then, were sentence completions, because, as Premack explains it, a question is just a statement with one or more elements missing. When Sarah was introduced to a new word, she was also presented with a sequence of familiar symbols, including a question mark, arranged on her board. Then she was handed the new symbol. It was the only unknown, and all she had to do, at first, was tuck it into the sequence at a location already pinpointed by the question mark. "One-to-one substitution may be the simplest of all training procedures," Premack said. When the unknown had been presented to Sarah often enough, as part of enough different sequences, she seemed to grasp the meaning, since now when she was confronted with the word in brand-new sentences, she responded appropriately.

Premack also set out to teach Sarah that things have names. She was given symbols for "name-of" and for "not-name-of," and she learned, working on a tray rather than on her magnetized board, to put the symbol for "name-of" between an actual apple, for example, and the plastic word for apple (so that her statement read, "'Apple' name of apple") and to put the symbol for "not-name-of" between the plastic word "banana" and the apple. Premack could now teach her a new word just by showing her an object, the symbol for "name-of," and the plastic word for the object.

There is something appealing about a method of teaching language that mixes tangible, three-dimensional objects with abstract symbols. As Premack puts it, this "bridges the chasm some philosophers interpose between words and things."

And yet words *are* at one remove from the objects they stand for: they're one step less tangible, and that's what makes language the powerful tool it is. Because words can substitute for the things they refer to, the name of an object can be used to call up from memory everything that is associated with the original object itself. For example, with Sarah, Premack could use either a real apple or the plastic word "apple" (a blue triangle) to elicit all the information she had about apples: she would write that an apple is red, not green; round, not square; and so on.

He could also use plastic symbols to teach her a new word, without actually showing her the object the new word represented. One day Sarah's teacher wrote on her board, "Brown is the color-of chocolate." All the words in the sentence were familiar to her except the word "brown," which was new. She definitely knew what the token for "chocolate" represented; however, there was no chocolate anywhere in her cage at the time. Next, she was shown four pieces of wood, each painted a different color, and instructed to pick out the one that was brown. She had no trouble doing it. Since she was never, in the course of the experiment, actually shown a piece of chocolate, she must have worked from a mental image of what chocolate looks like. In other words, she had her own internal symbol for chocolate.

It's the ability to symbolize that makes language possible. Since Sarah obviously could symbolize, either she learned to do this in the course of learning how to read and write, or else it was an ability she already had—which is what Premack believes. He speculates that symbolization may be part of all learning. In that case, it must be widespread among animals. It's interesting to consider what forms it might take. When a dog, longing to go for a walk, noses its leash where it hangs in a hallway, what is going on in the dog's head? Does the leash call up mental pictures of grass and trees? Does it revive memories of the scent of squirrels or of the dog next door? We don't know the answers now, but Premack believes that we will have them someday, if we can find a way to teach words to a dog, as we are now teaching them to chimps.

We know that chimps respond to symbols, but how much further can they go? Or, as Premack puts it, how many transformations can one make on an object and still have the animal recognize it? We might, for example, go from actual action to a motion picture of the same action to the same film speeded up—and right to the end, a chimp would recognize what was happening. It seems that apes can travel a long way from three-dimensional reality and still make sense of what they see. We know they like to watch television and look at pictures in magazines. (Dogs, on the other hand, seem indifferent to both.) Chimps also recognize their own image in a mirror, something only apes can do: anesthetize a chimp, paint a blue spot on his ear, and then when he wakes up, show him a mirror, and his hand will go instantly to his ear. Monkeys, put through the same procedure, seem to think it's some other animal looking back at them from the mirror.

But, to come back to the original question, how far *can* chimps go—how many transformations can they follow? Premack once presented Sarah with a whole apple, an apple that had been quartered, and three photographs. The photos showed a crayon, a bowl of water, and a knife, and Sarah was required to choose which of the three formed a logical connection between the whole apple and the cut apple. She chose the knife. This, Premack explains, is even more impressive than an ability to make sense of a speeded-up film, for no continuous sequence was presented. Sarah didn't see the apple splitting below the knife; all she saw were the end points, a stylized representation. "Probably no other animal except humans and apes can recognize a representation of its own behavior like that," Premack said. "Any mind that can do that can learn language to some degree. In fact, that's trivial, given the chimp's excellent memory."

The part memory plays in language is often underrated and sometimes completely overlooked. Professor Premack points out that it may be memory, more than anything else, that makes language possible. Actually he believes that many different facets of intelligence play a part in linguistic ability. Memory is one; syntax, another; and a third factor is the way the individual perceives the world.

Language is mapped onto our perceptual world, Premack says, and humans see the world in terms of cause and effect. This fact is reflected in every conceivable aspect of language—we can analyze almost any statement in terms of who does what to whom; of agent, action, and object. However, once again, humans are not the only animals able to analyze events this way, because any creature smart enough not to bump into the same obstacle twice must in some way be operating in terms of cause and effect.

As for syntax, it doesn't come easily to chimps. They are a lot slower at learning it than human children are. However, Premack *was* able to teach Sarah to do some fairly complex things. She learned to take singular nouns and make them plural. She learned to follow instructions when her trainer wrote that she was to put "green on red" or "red on green," an important lesson since she had to rely on word order for the meaning. Eventually she learned, reluctantly, to use the copulas "is" and "is not." Some languages, including Russian and ASL, don't have copulas. Children learning English learn them late. Apparently they're either difficult to master, an unnecessary frill, or perhaps both.

One of Sarah's most impressive achievements was the ability to understand the conditional. For this lesson, two sentences were spelled out side by side on her board: "If Sarah take apple, then Mary give Sarah chocolate," and "If Sarah take banana, then Mary no give Sarah chocolate." Sarah learned, though with a great deal of difficulty, to select the apple rather than the banana to earn her chocolate; and she was able to transfer her skill when other words were substituted—to perform correctly, for example, when the instructions read, "If Mary take red, then Sarah take apple."

Once again, however, *conditional discrimination* (which is

what Sarah was doing in this lesson) is not a skill used only in language. Psychologists have long since demonstrated it in other situations. A chimp can be taught that in any set of three, she is to pick the oddity. Given two greens and a red, for example, she is to choose red; given two circles and a square, she is to select the square. Then the conditional can be added: the chimp is presented with three objects—a blue triangle, a blue circle, and a yellow circle—and she is taught that *if* the objects come on an orange tray, *then* she is to choose by color, picking as the oddity the yellow circle; but *if* the objects arrive on a green tray, *then* she is to choose by shape, selecting the triangle and leaving the circles. Even a rhesus monkey can do this trick. In learning to read a conditional sentence, then, Sarah was once again using an ability chimps already have.

Under Premack's tutelage, Sarah's language ability developed steadily. On most tests of her skills, she scored about eighty percent. However, eventually he had to face the Clever Hans problem.

Clever Hans is the bane of the experimental psychologist's existence. He was a horse who became famous back at the turn of the century for his mathematical abilities. Presented with an arithmetic problem, Hans would tap out the answer with his hoof. Because his owner, an elderly Berlin schoolteacher, didn't exhibit him for money and sincerely believed in his ability, and because Hans could perform even when his owner wasn't present, the horse was taken seriously, and in 1904, a scientific commission was appointed to look into the matter. The scientists soon figured out that Hans was being cued by his human audience through subtle, unintentional body movements. When the math problem was first presented, spectators would lean forward slightly in anticipation; at that point, the observant Hans began to tap. As he reached the correct answer, the watching humans would change posture slightly and Hans would stop. The investigators found that when the people around him didn't know the answer to the problem, neither did Hans. So Hans was discredited and his owner died soon afterward—of heartbreak, or so the rumor went. In recent years, psychologists have proved that humans often subconsciously pick up the same kinds of unintended cues: in psychological studies, the experimenter often gets the results he expects to get because he unwittingly telegraphs his expectations to his subjects.

To get around the Clever Hans problem, the Gardners designed a special apparatus for testing Washoe. The experimenter in the room with her couldn't see the slides or objects she was being asked to identify, so he couldn't cue her. With Premack's system, it was harder to rule out unintended cueing, and Sarah's trainer could conceivably have been tipping her off without ever meaning to by something as simple as an occasional swift glance at the plastic word that was the correct answer. Premack eventually ran a special test for Sarah, using a trainer who didn't understand her language. This young man, John, through earphones and microphone, was in constant verbal contact with one of Sarah's regular teachers, who was stationed out of sight outside the room. The symbols were all numbered, and John was given lists of numbers to go by as he arranged Sarah's plastic words in sequences on her board. He reported to the trainer outside the numbers on the words Sarah selected for her answers, and was told whether to reward her or not. In this test, Sarah didn't perform as well as usual, though she did do considerably better than she would have been expected to do if she were just guessing. Premack speculates that she may have been disturbed because she was in a novel situation and cut off from her usual trainers. He points out that a child being tested by a total stranger often doesn't do very well, either, and that on the Gardners' tests Washoe was only 55% correct.

Professor Premack's work has already born practical fruit. Variations on his system have been used in many places to teach language to speechless humans—to retarded children and to adults who have lost the ability to speak because of brain damage. Dr. Joseph Carrier, a speech pathologist who is now at the University of Kansas Bureau of Child Research, used his own adaptation of Premack's plastic language to work with children who were totally languageless or who knew only one or two words. Of 180 youngsters taught by the Carrier method during a two-year period, only two failed completely. At the end of that time, 53 were still in training; but 125 had learned to communicate with symbols, and these children had all gone on to regular speech therapy because, typically, after a number of lessons with the plastic symbols, the children would spontaneously begin to make sounds—to attempt to talk.

Why is it that it's easier to learn to communicate with plastic symbols—or with signs—than it is to learn to speak? You will recall that chimpanzee infants begin to sign when they are just three or four months old, and that human children often begin signing at the age of five months. Babies are, then, obviously capable of language long before they're capable of speech. In fact, Mary Morgan, Sarah's principle trainer, easily taught her own ten-month-old daughter to ask for certain baby foods by using colored blocks that functioned as names. The baby dropped the appropriate block down a special chute and was given the food she requested.

It seems that speech itself is much more difficult than most people think it is. There are some two hundred different muscles in the lips, tongue, throat, and abdomen that can be involved in speech. To utter words, one must use the right muscles and coordinate them correctly, and furthermore, one must be able to remember which sounds to make, and in what order, to produce the intended words. As Carrier put it in a *New York Times* interview, for the retarded child, producing speech may be the equivalent of "rubbing his stomach with one hand, patting his head with the other, tapping his toes, and blinking his eyes, all at the same time."

When Premack began his work with Sarah, he didn't really anticipate that it would have any practical applications, though he's delighted that it does. Meanwhile he has moved on to new projects. He was forced to retire Sarah temporarily, back in 1971 after about four years of training. "She's a brat," he said, "very temperamental and spoiled and given to tantrums."

So he began again with two younger chimps, Elizabeth and Peony. Elizabeth seemed to enjoy her language lessons; Peony did not and was, in fact, a much poorer student. Yet under Premack's system, they both learned at about the same rate. When I interviewed Professor Premack in the summer of 1975, Peony knew seventy-five words and at the age of seven was still docile enough to work with. "She was reared differently," he told me. "Had we been less amateurs with Sarah . . . " He let the sentence trail off regretfully. Elizabeth, unfortunately, had died, but at the time of her death she, too, knew seventy-five words.

Sarah and Peony both went along with David Premack when, in the fall of 1975, he moved to Philadelphia to become a professor at the University of Pennsylvania. There he has a sizable compound for his chimps—about two-thirds of an acre—and a language lab that the animals can wander into at will. At this writing in 1977, four young chimps are already in residence, with two more expected; all will eventually be taught the artificial language.

Sarah has her own quarters in the lab, though she's not allowed into the compound with the other animals, and Premack is working with her again. He has a new device, a kind of teleprinter with two keyboards, one for Sarah and one for an experimenter to use outside her cage. When either Sarah or her trainer operates the keyboards, the familiar symbols appear on a videoscreen for both to see. When I last spoke to him, Premack was obviously delighted to be able to continue Sarah's training.

"She may be a dangerous animal," he said, "but she is very, very bright."

6.

The Lana Project

He was nearly ten before he commenced to realize that a great difference existed between himself and his fellows. His little body, burned brown by exposure, suddenly caused him feelings of intense shame, for he realized that it was entirely hairless, like some low snake, or other reptile. . . .

And the little pinched nose of his; so thin was it that it looked half starved. He turned red as he compared it with the beautiful broad nostrils of his companion. Such a generous nose! Why, it spread half across his face! It certainly must be fine to be so handsome, thought poor little Tarzan.

—TARZAN OF THE APES,
by Edgar Rice Burroughs

In Atlanta, Georgia, another artificial-language experiment is taking place in a setting that's downright surrealistic.

Imagine a clear plexiglass cube seven feet high and seven feet wide. Inside this square fishbowl crouches a short, hairy, humanoid creature, a chimp named Lana. At the moment, Lana is confronting a wall console, a keyboard made up of row upon row of rectangular plastic push buttons. There are over a hundred buttons in all, and each glows with its own dim light: red, blue, violet, black.

Thoughtfully scanning the console, Lana reaches up with one long arm and casually pulls down an overhead lever. Then she punches four buttons in rapid succession. As her finger touches each button, it lights up instantly to neon intensity. Simultaneously, four symbols appear on a row of screens above the keyboard, and in a control room outside the plexiglass cube, a teleprinter types out a message: "Please machine give M&M." Lana has made a request. Inside the cube now, a tone sounds approvingly, indicating that she has produced a grammatical sentence. At the same moment, a handful of M&M candies rattles down into a slot below the keyboard.

Since December, 1972, Lana has used this exotic, computerized vending machine to control the world around her. She orders up food and drinks, asks to have the window opened, requests slides, music, or a movie (such as *Primate Growth and Development—A Gorilla's First Year*). She also uses her keyboard to hold conversations with her trainers.

The Lana project, located at the Yerkes Primate Research Center, is directed by Professor Duane M. Rumbaugh, chairman of the psychology department at Georgia State University. I interviewed Rumbaugh in 1975, after a lecture he gave in New York. He explained that Lana's language (it's called Yerkish) was designed to be as unambiguous as possible, with no fuzzy constructions and no words with several alternative meanings. In addition, her machine will respond only to sentences that are grammatically correct. Each key on her console has a symbol, or lexigram, on it and represents a word. The background colors are a clue to the category the word falls into. Red keys are for food and drink, black are for prepositions, violet are for animate beings—for people, animals, or Lana's machine—and so on.

Her training began when she was about two years old. At first, all she had to do was press one key to get what she wanted, hitting just the "M&M" button, for example. Once she had learned to do this, the task was made harder. Now she had to press three keys to get the same results: "Please M&M period." The "please" and the "period" weren't added out of any desire to teach her manners or punctuation; they were necessary to signal the computer to expect a request and to inform it that the message was over.

Next, the middle key had to be replaced with a phrase, but the words of the phrase were at first tied together, so that if Lana pressed any key that was part of it, the others would also light up and their symbols appear on the display screens . For example, to write "Please (machine give M&M) period" all she had to do was to punch "please," "give" (or "machine" or "M&M") and "period." After that, the phrase was gradually broken down into its individual words. In the end, Lana was typing out whole sentences word by word. New lexigrams, as they were added to her board, were moved around frequently to make sure she learned the lexigram itself and not just its position.

At first, Lana shared her quarters with an orangutan named Biji, but he proved to be too much of a distraction from the lessons, so he was moved out. However, she still had some contact with other apes when she was taken outdoors to play. She was well adjusted and gentle with the technicians who worked with her, and when she did misbehave, the most effective way to punish her was to disconnect her keyboard temporarily.

Lana is said to be a rather average chimp, no brighter and no dumber than any of the other baby chimps born at Yerkes. Yet she proved to be a much quicker pupil than her teachers had anticipated. In fact, by the end of the first six months, she was further ahead with her lessons than her human mentors had originally thought she might ever be. She had, for example, figured out that if she made a mistake, there was no point in finishing the sentence. She learned that if she hit the period key, this would, in effect, erase what she had written, so that she could begin again immediately. Though it hadn't occurred to her trainers to give her a way to erase, she discovered one.

Her teachers had also not been at all sure that she would learn, on her own, to read the symbols displayed on the screens above the console. However, before long, they realized that that was exactly what she was doing, since several times she accidentally hit the "please" key with her foot while climbing on the overhead bar. Discovering later that "please" was already displayed, she simply filled in the rest of a request.

To test Lana's reading ability further, Rumbaugh designed a sentence-completion exam. A human companion would begin a sentence, using a keyboard outside Lana's room; the symbols appeared on her display panel, and she was left to complete the sequence. Some of the sentence fragments the human supplied were a valid way to begin a sentence ("Please machine give . . .") and some were not ("Please piece give . . ." "Please machine Tim . . ."). Lana did remarkably well: 94 percent of the time, she correctly completed the valid beginnings, and 90 percent of the time, she promptly erased the invalid ones by pressing the "period" key.

The list of Lana's accomplishments is long and overlaps what Sarah and the signing chimps can do. She uses pronouns and prepositions appropriately, understands a great deal of spoken English, sometimes asks herself questions and answers them, and often initiates conversations with her trainers. Her use of language is quite flexible—in asking for coffee, she has phrased her request twenty-three different ways—and, what is perhaps most important, she has combined words into new sentences that she was never taught. For example, in the small hours of the morning, when no humans are around, she is apt to type out wistfully, "Please machine tickle Lana," or "Please machine come into room." She has also invented names, dubbing an orange "apple which is orange," for example, and a cucumber "banana which is green." And she has generalized: taught the word "no" as a simple negative (meaning "it is not true that . . ."), she soon began to use it also as a protest. One day she saw a technician drinking Coke outside her room and there was none available in her machine; she stamped her foot in frustration and thumped the "no" key hard.

But Lana's greatest breakthrough came on the day she finally asked for the name of an object. The idea that everything has a name and that names are useful is perhaps the crux of language. Anyone who has read Helen Keller's autobiography or who saw the play *The Miracle Worker* will remember the scene in which Helen's teacher held her hand under running water and, using sign language, spelled into her other palm the word "water." Helen was blind, deaf, and, at the time, totally without language. For weeks, her teacher had been spelling into her hand, and she had learned to associate different finger movements with different objects, but it was all just a game to her until that day by the pump, when suddenly she realized that the movements she felt against her palm were a way to identify something and so to reach out to the world. Dropping to her knees, Helen asked for the name of the ground, the name of the pump, the name of every object she touched. Within just a few hours, she learned thirty words.

Lana's discovery of the usefulness of names was, of course, not quite so dramatic. In the beginning, it was difficult to get across to her the idea that things have names. It took two weeks and sixteen hundred tries before she began to use the lexigram for "name-of" correctly. Then one day Timothy Gill, her trainer, entered her room carrying a bowl, a can, and a box. Lana knew the symbols for "bowl" and "can" but not for "box," and it was the box that was baited with candy. With both Tim and Lana using Lana's console to communicate, this is the conversation that took place:

LANA: ?Tim give Lana this can.
TIM: Yes. *(He gives her the empty can.)*
LANA: *(Putting the can aside.)* ?Tim give Lana this can.
TIM: No can.
LANA: ?Tim give Lana this bowl.
TIM: Yes.
LANA: *(Putting bowl aside.)* ?Shelley.
TIM: No Shelley. *(Shelley, another trainer, was not around at the time.)*
LANA: ?Tim give Lana this bowl. *(Before Tim can answer, Lana types out another sentence.)* ?Tim give Lana name-of this.
TIM: Box name-of this.
LANA: Yes. ?Tim give Lana this box.
TIM: Yes. *(She rips it open and takes out the candy.)*

Later the same day, when Tim brought a cup into her room, Lana asked for the name of it.

Time and again as the project progressed, Lana startled her trainers by her creative use of language. There was the day her machine broke down and failed to deliver bread, though she asked for it repeatedly. Her solution was to type out a message to a technician who happened to be nearby: "Beverly move behind room period." She had asked humans to go behind the room before, but only as part of a game they played. Once they were in position, she would ask the machine to open the window and then the technician would perform for her, blowing smoke rings or bubbles or some such thing. However, access to the loading end of her machine was also behind the room, and, sure enough, once Beverly went back there, she realized that the bread-vending device wasn't working and fixed it. To make certain Lana hadn't simply hit on the right response by chance, equipment failures were deliberately staged several times after that; each time she asked the technician to move behind the room.

Lana's ability to construct new sentences and use them to cope with unfamiliar situations was also impressive. At one point, Timothy Gill, trying to find out what it would take to get a good, long conversation going with her, designed a set of experiments in which he deliberately frustrated her, since he believed that conversation is primarily a problem-solving activity. Late one afternoon, at a time of day when Lana normally got her ration of Monkey Chow, Tim loaded her machine with water instead. Once she discovered what he had done, she asked for chow. He lied, saying there was chow in the machine. Then she asked for juice, and the following exchange took place:

TIM: Juice in machine. *(Once again, lying.)*

LANA: No juice. *(Here she was apparently declaring that Tim's statement was not true, the first time she had ever made such a comment.)*

TIM: ?What in machine.

LANA: Water in . . . milk. *(She hesitated over various keys such as "machine" and "room" before pressing "milk." She seemed to be saying, as best she could, that water had been loaded in the place where her milk normally was. This was also a novel statement.)*

LANA: ?You want water in machine. *(Perhaps she wanted to know whether Tim really meant that water should be there. Also a novel statement.)*

TIM: No.

LANA: ?You move water behind room.

TIM: Water behind room. *(This was true.)*

LANA: You move water out-of machine. *(A new and appropriate request. She had only asked a very few times before to have something taken out of the machine.)*

As of March, 1977, the last time I talked to Professor Rumbaugh by phone, Lana knew—and had on her keyboard—109 lexigrams, and her language, like Sarah's, was being used to teach speechless humans. Her computer system had been moved, virtually intact, from Yerkes to the Georgia Retardation Center, where severely retarded children are now learning to type out sentences such as "Please experimenter give popcorn period." A more appropriate vocabulary has been substituted for Lana's, and different training techniques are being tested; the hope is that eventually the children will graduate to spoken language.

Lana herself has had a new and more elaborate system installed—where the old computer could only process sentences up to six words long, the new one can handle ten words—and five young chimps have been added to the language training program. Before long, Lana will be able to communicate with them through her keyboard. In addition, a portable unit called a conversation board has been designed, so that Lana and her "machine" are finally mobile. The unit is the size of a small suitcase and weighs about fourteen pounds. It can be carried in a shoulder sling or backpack and is powered by batteries. There are 112 keys, but as Lana's vocabulary grows, new sets of lexigrams can be substituted to suit particular situations. It's Rumbaugh's guess that Lana is perfectly capable of learning four hundred or five hundred lexigrams if he decides to teach her that many. Like Premack and the people who work with signing apes, Rumbaugh's main focus is not on vocabulary but syntax, and certainly Lana has already demonstrated that chimps can be taught simple syntax.

In the future, Professor Rumbaugh is planning to give Lana names for the numbers from one to ten and to see whether she can manage simple arithmetic. He already knows that, shown two trays, each with a different number of small objects on it (for example, five washers of various sizes on one tray and two on the other), she can say which has "more" and which "less." This may mean that she can count and perhaps even do simple addition.

Still further into the future, Rumbaugh would like someday to use Lana as an interpreter, to explain to humans why chimps do the things they do in their own communities.

Despite Lana's impressive performance and the accomplishments of Sarah and of the signing apes, there are still any number of linguists who heatedly deny that any ape has achieved language.

Part of the problem is that there is no generally accepted definition of "language"; nor is anyone ready to say that at some particular point in their development, children have achieved language, whereas before that, they hadn't. It's simply assumed that children, from the time they utter their first words, are producing "language" (whatever that is) because they grow up to be adults who speak or sign. Yet no one will extend the same courtesy to a chimp and say that Washoe, for example, had language from the time she formed her first sign.

Instead, some linguists keep redefining language, pushing the boundaries back so that they will continue to exclude apes. What is really characteristic of language, these people say now, is that it's governed by principles of great formal complexity. What chimps can do, in terms of syntax, is trivial in comparison with what humans can do; therefore, chimps don't have language.

On the other hand, the scientists who have been teaching language to apes believe that we must stop thinking of language as a yes-or-no proposition, an ability a species either has or hasn't got. Instead, we must begin to see language skills as a continuum and to admit that a chimp's skills do fall somewhere on the continuum and are definitely related to the skills of a human being.

Some scientists are beginning to adopt an even broader perspective, to look at language—ape and human—in the light of the whole wide spectrum of the way animals communicate. As the perspective expands, it seems time to ask: if we have underestimated apes, have we done the same with other animals?

Certainly it has been the scientific view up to now that the signaling of animals is totally unlike human language. Scientists have told us that with many of the lower animals, communication is virtually automated. One animal produces a stimulus; another reacts with a response; and both are simply going through motions programmed in their genes. The courtship of crickets and the aggressive displays of Siamese fighting fish seem as inexorably predetermined as the fact that a fish has fins and a cricket doesn't. More intelligent animals, the scientists explain, can express their emotions and even their intentions: one dog can threaten another, which may in turn threaten back. However, they can't confide past histories or plan for the future, so how can one possibly compare what they do with intelligent conversation?

Perhaps one can't. But, on the other hand, a close look at animal signaling suggests that some species, at least, are able to communicate a great deal more than we once thought they could—for example, one school of thought holds that teaching a chimp to sign or to write may not improve much on its natural ability to communicate with gestures.

Let's try to look at communication, then, as one long continuum, stretching from the apparently automated signaling of the simpler species to the Gettysburg Address. Let's ask what animals communicate about, and why and how, and then analyze human exchanges along the same lines. Then we'll be ready to reconsider some of the questions raised in these opening chapters—questions such as:

- What is language?
- What, if anything, are we doing to or for apes when we give them words?
- And where are we all going from here?

7.
Slime Molds: The Messages They Send

The uniformity of the earth's life, more astonishing than its diversity, is accountable by the high probability that we derived, originally, from some single cell, fertilized in a bolt of lightning as the earth cooled. It is from the progeny of this parent cell that we take our looks; we still share genes around, and the resemblance of the enzymes of grasses to those of whales is a family resemblance.

—THE LIVES OF A CELL,
by Lewis Thomas

On the screen, they were simply little white blobs, meandering about and occasionally bumping into one another, rather like a New Year's Eve crowd in Times Square. Actually they were slime-mold amoebas, microscopic one-celled animals magnified many times and moving at superamoebic speeds thanks to time-lapse photography, which showed them hurrying along at five hundred times their normal pace. Wiggling, constantly changing shape, and constantly moving, they were making short work of a field of bacteria—their normal diet—which looked like pinpricks in the background.

"If you watch closely, you'll see mitosis," Professor John Bonner told me. "There!" He used the shadow of his finger to point and, sure enough, one of the white blobs split across the middle and became two, reproducing by fission in the ancient way of some one-celled animals.

Professor Bonner, chairman of Princeton University's biology department, has spent his professional life studying the cellular slime mold, a slippery dab of primordial ooze that lives in the soil and is like a fungus in some ways, like an animal in others. I had come to Princeton to talk to him about slime molds because it seemed to me that it would be worth finding out how lower forms of life communicate. I wondered: what are the simplest, the most indispensable messages that animals send?

I was also, frankly, hooked on slime molds in advance because of what I'd read about their remarkable life cycle, for they switch identities more often than Merlyn the Magician. They start as individual one-celled animals, sturdy individualists, but then in times of famine, they aggregate, coming together to form a kind of slug, a composite animal that creeps through cavities in the soil. Eventually the slug, in turn, converts itself into something like a plant, a long-stemmed beauty topped with a globe of spores. These are then scattered abroad in the way of plants and become solitary one-celled amoebas again.

Through the medium of time-lapse photography, Professor Bonner showed me all these life stages. Scene one had caught the individual amoebas feasting on a field of bacteria. Next we came to scene two, in which a batch of amoebas, facing starvation with no bacteria available to feed on, began to clump together. At first, they seemed simply to coagulate, but soon they were streaming in rivulets toward a central collecting place. From my reading, I knew that the outlying animals were being called in by chemical signals. At regular intervals, the amoebas at the center were giving off puffs of a chemical called an *acrasin* (pronounced *a-cray-zin*). Catching a whiff of it, animals nearby began to move toward the source of the signal and at the same time produced puffs of acrasin of their own. In this way, the signal was relayed from animal to animal, out to the fringes of the group.

Scene three: the aggregating amoebas disappeared, and in their place were several slugs snaking their way across the screen. Though they were only a millimeter or two long, barely big enough in nature to see with the naked eye, each was composed of as many as 100,000 individual amoebas literally stuck together and moving like a single animal, drawn toward warmth, toward light. As each slug oozed along, it secreted a kind of overcoat of slime to use as a tunnel—as something to glide through and then abandon.

Scene four: a migrating slug bunched up, momentarily taking on the shape of a chocolate drop; then it began to thrust up into the air, climbing sinuously like Jack's fast-growing beanstalk and carrying aloft at its tip that plump droplet of spores. Though I couldn't see it on the screen, I knew that the amoebas at the front end of the slug had become the stalk of this so-called fruiting body. Trapped inside a cellulose outer sheath that they had constructed for themselves, ultimately they would die. The animals from the tail of the slug had become spores, each encased in its own hard capsule.

The film was over, and as he turned on the lights, Professor Bonner filled me in on the rest of the slime mold's life story. In the wild, he said, sooner or later a worm or some other insect presumably comes zipping by. The spores stick to it and are dispersed. Some land in a favorable place, and a few hours later the outer case splits open. A single amoeba hatches and sallies forth to feed on soil bacteria—until the next famine, when it will be time to aggregate again.

How on earth does one study the chemical signals of microscopic animals, and how does a man come to devote his professional life to the contemplation of a creature so minuscule? (*Amoeba proteus*, the animal generally studied in biology courses, is forty times larger.) When I asked Professor Bonner these questions, he explained that originally he was interested in embryology. However, embryos are difficult to study because their cells grow as they differentiate, whereas the slime mold, in turning into a fruiting body, differentiates without growing. So research on slime molds may someday help explain how embryos develop.

The cellular slime mold is there, invisibly, in all its life stages, in almost any spoonful of garden dirt. The individual amoebas are similar to leukocytes, the white cells in human blood, and they consume soil bacteria much as leukocytes consume bacteria in the bloodstream. Their existence was first noted by humans in

1869. Today about thirty-five different species are known to exist, and the number keeps rising. To date, *Dictystelium discoideum* (known familiarly in the lab as D.d.) has been the species most studied.

The slime mold is probably the most sociable of the one-celled animals, and the fact that it sends out chemical signals while aggregating has been known for quite some time. Recently, however, scientists have learned that it has signals to suit at least half a dozen other situations as well.

It has, for example, several different ways of saying "Keep away," for in almost all of its life stages—as a solitary hunter, as an aggregating center, and as a fruiting body—the slime mold produces repellents that force other amoebas (or centers, or fruiting bodies) to keep their distance. Presumably these spacing mechanisms were designed by evolution to insure that the animals spread out and don't simply decimate the bacteria in one tiny area and then starve to death, or, in the case of the fruiting bodies, to insure that spores are widely dispersed.

Then there is the acrasin. It was, in fact, Professor Bonner who originally coined the term *acrasin*—after the witch Acrasia in Edmund Spenser's *Faerie Queene,* who attracted men and transformed them into beasts. For years, scientists struggled to prove that such a chemical existed, as they knew it must, but it was somehow too ephemeral to capture and analyze. In fact, when researchers slipped a freshly killed D.d. center in among a crowd of hungry D.d. amoebas, they stubbornly refused to aggregate, though it seemed that the odor of the acrasin should still have been clinging about the center. Finally Dr. Brian M. Shaffer of the University of Cambridge solved the mystery by demonstrating that aggregating amoebas actually puff out two chemicals, an acrasin and a companion enzyme that rapidly deactivates it. Without this canceling substance, so much of the acrasin would accumulate during aggregation that it could never serve as a direction signal.

Dr. Shaffer's experiment involved what Bonner has called "his Rube Goldberg demonstration." He suspended an ordinary sausage casing inside a glass container, painted the outside of the casing with amoebas that were ready to aggregate, and then dripped water along the inside of it. The small molecules of the acrasin were able to pass through the casing and become mixed with the water, while the larger molecules of the destructive enzyme were blocked out. Thus the water that collected at the bottom of the container contained stable acrasin, without the canceling substance. When drops of it were placed in the neighborhood of hungry amoebas, they aggregated nicely.

The slime mold also produces another set of chemical signals to control differentiation: they cause the amoebas at the front end of the migrating slug to become stalk cells in the fruiting body and those in the back end to become spores. For the amoebas initially sort themselves out, with prestalk cells at the front of the slug and prespores at the rear; but all the same every individual little animal has the potential to become either stalk or spore and, in fact, can switch identities up until the last moment. Scientists know this is true because they've found that they can chop off the front end of a slug and it will grow into a small but perfectly proportioned fruiting body. In the whole snipped-off column of animals, all destined originally for stalkdom, an appropriate number get the message that now, instead, they are to turn into spores.

How do we know that it's the front end of the slug that develops into stalk while the back converts to spores? The proof comes from a study done by Professor Kenneth Raper, who fed D.d. amoebas red bacteria until they, too, turned red. Then he chopped up migrating slugs and grafted red front ends onto the back ends of normal-colored slugs (normally they're transparent). The result: fruiting bodies with red stalks and transparent spores.

When I asked Professor Bonner how one grafts amoebas, he explained, "Usually we just slice the slug like a sausage and then rearrange the bits. We use a tiny knife; or else I might pluck out one of my own eyelashes, since it's nicely tapered, fairly stiff, and of course quite handy. There's a newer method, too, where you use a hypodermic to inject one lot of cells into another lot."

Under a microscope in his lab, he showed me slugs that had been colored in a different way, dyed with a chemical that affects prestalk and prespore cells differently. Red to the midpoint, they had tail sections as transparent as Vaseline, and each towed a jet trail of slime. In another part of the same shallow glass case—called a petri dish—a fruiting body loomed like an exotic red cactus above a scattering of red boulders—aggregation centers that had not yet converted into slugs. It was hard to believe that each was a whole society of living, breathing, communicating creatures.

Besides the repellents, the acrasin, and the chemicals that control differentiation, slime molds also have at least one sexual signal in their repertoire, for it was recently discovered that they don't just reproduce by fission; they have a sexual stage as well. The D.d. colonies that are studied in labs like Bonner's are sometimes culled from local soil, but more often they're laboratory-grown descendants of two original stocks, which are labeled NC 4 and V 12. Thought to be different strains of D.d., for decades NC 4 and V 12 were kept in separate cultures and used, generation after generation, for lab studies—until one day someone discovered that they were actually opposite mating types.

When I asked Professor Bonner which is male and which female, he explained that with lower forms of life, "We never know quite what we mean by male and female, so we speak of plus and minus strains." Sex in the slime mold is, in fact, altogether a curious phenomenon. The amoebas collect in one place, much as in aggregation, but once they've lumped together, they grow a protective surrounding wall. Then one cell eats up all the others. Inside this bloated cannibal, the chromosomes of the victims pair up and genes are exchanged—the genetic dice are shaken. Eventually the cannibal begins to divide into many small cells again, the walls rupture, and a new generation of amoebas emerges.

Professor Bonner explained that it's still not clear exactly how sex fits into the life cycle of the slime mold. Sometimes when they're mixed together, the plus and minus strains refuse to form a cyst; sometimes they refuse to do anything else. It's thought that probably they encyst only during hard times—for example, when there is too much water around. It's almost as if they need to grow that wall to avoid being drowned. However, evolution has also seized the opportunity for an exchange of genes, since without a way to recombine genes there can be no evolution—no chance to come up with a maverick offspring who just happens to be equipped to survive in changed circumstances.

Miloslava Machac, a student of Professor Bonner's, demonstrated that at least one sex hormone exists to bring the plus and minus strains together: she used a piece of dialysis membrane to separate a crowd of NC 4s from some V 12s, and cysts formed on one side of it. The amoebas themselves were too large to cross through the membrane, but one group apparently projected a hormone through it that persuaded the other group to build cysts. Bonner believes that there are probably at least two hormones involved, that each strain must be able to attract and affect the other. There may actually be a whole series of hormones, as there are in other lower forms of life, in which one chemical triggers another, which triggers another, and so on, for a domino effect.

Once science has proved that a chemical signal exists, the next step is often to try to identify and even synthesize it. This has now been done with the D.d. acrasin. In the beginning, researchers tested a number of natural substances produced by higher animals to see if they would persuade D.d. amoebas to aggregate, but nothing worked. Finally, on a hunch, two scientists working at Princeton tried cyclic AMP, which is implicated in the action of hormones in many mammals, including humans. The amoebas responded beautifully. It seems that, at least in D.d., the acrasin *is* cyclic AMP.

There are fascinating similarities and dissimilarities in the roles that cyclic AMP plays in the life of D.d. and in the human body. For one thing, in humans, it's broken down by an enzyme very similar to the one that deactivates the acrasin in D.d. For another, just as D.d. uses cyclic AMP to carry messages between distant animals, so human chemistry uses hormones, with cyclic AMP employed as a secondary messenger at the site of the action. Assume that a human panics: adrenaline is pumped into the bloodstream. Carried to the liver, it triggers cyclic AMP that's trapped within the liver cells; this in turn triggers a cascade of enzymes, each sparking the next until energizing glucose is finally released from the cells into the bloodstream. Cyclic AMP also mediates between many other hormones and their ultimate effects. It is, in fact, such a fundamental part of the human hormone system that it seems possible it may have been invented first, with the hormones themselves added as an evolutionary afterthought.

In the world of the slime mold, cyclic AMP is ubiquitous: all the different species seem to use it during differentiation to generate stalk cells (presumably some other chemical produces spores). Soil bacteria also secrete cyclic AMP, and it's reasonable to assume that slime molds—which can sense their prey at a distance—are tipped off by this and other chemicals the bacteria produce. In fact, evolution probably seized on this sensitivity in creating the signal system that D.d. uses during aggregation.

However, not all slime molds use cyclic AMP as an acrasin. (Remember, there are about 35 different species—D.d. is only one.) Actually, acrasins are species-specific—which helps prevent cross-breeding—and one of Professor Bonner's present projects is to try to identify the one used by the species called *Polysphondelium violaceum.* He believes its acrasin may be a peptide, and again there is a connection to human chemistry, since many hormones in the human body are peptides.

All this is intriguing from an evolutionary point of view, because of course all multicelled organisms originally evolved from primitive, single-celled creatures. There are several different theories about how this happened, one of which suggests that the origin was a colonial "animal" created by aggregation. If that's true, the key to survival must have been excellent internal and external signal systems.

Slime molds are, then, basic to the story of communication, though not quite in the way I expected. It's not really a case of simple animals exchanging simple signals. Instead, slime-mold communication seems to be distantly related to the internal communication systems that allow the rest of us to move around, fend one another off, and occasionally to aggregate.

8.

Strictly Between Fish

The line rose slowly and steadily and then the surface of the ocean bulged ahead of the boat and the fish came out. He came out unendingly and water poured from his sides. He was bright in the sun and his head and back were dark purple and in the sun the stripes on his sides showed wide and a light lavender. His sword was as long as a baseball bat and tapered like a rapier and he rose his full length from the water and then re-entered it, smoothly, like a diver and the old man saw the great scythe-blade of his tail go under and the line commenced to race out. . . .

He is a great fish and I must convince him, he thought. . . . thank God, they are not as intelligent as we who kill them; although they are more noble and more able.

—THE OLD MAN AND THE SEA,
by Ernest Hemingway

"Why is it," I asked George Barlow, "that nothing interesting ever happens in my fish tank? I assumed that once I'd done some reading about fish, I'd be able to see mine communicating. They do threaten each other occasionally, but aside from that, I haven't been able to catch them in the act."

"You know," Professor Barlow said, "the less an animal is like a human, the less we're able to hook into its system of communication. I've often tried to explain to visitors what's going on in there." He gestured at the aquarium on the opposite wall of his office. "Usually they look at me as if I'm crazy—they can't see what I see. But in just a week or two, I can teach undergraduates what to look for; after that, they start telling me what I've missed."

George Barlow, according to zoologists I've talked to, is perhaps the world's foremost authority on cichlid fish. A zoology professor at the University of California at Berkeley, he's a tall, soft-spoken man with curly gray hair. His particular animal at the moment is the Midas cichlid, *Cichlasoma citrinellum,* and he has donned scuba gear to observe it where it lives in the depths of lakes in Nicaragua and has also studied it in his lab at the university.

Surprisingly, fieldwork on fish is relatively rare—it seems underwater observation is difficult. Professor Barlow explained that scuba equipment makes a lot of noise, and while some fish don't seem to be disturbed by this, others simply disappear. The scientist's vision is also limited underwater, and furthermore, if he or she wants to study fish that live more than thirty feet down, diving is limited to about three hours a day (someone who is careless and stays down longer is apt to get the bends). In some ways, it's actually easier to study chimps in the jungle than cichlids in the murky depths of a lake.

Professor Barlow once had the chance to study reef fish while living in an underwater habitat. Since a habitat is actually a decompression chamber, there was no limit to the amount of time he could spend in the water, and he could follow one particular fish for quite a while, noting down what it did, using an ordinary pencil on a white plastic slate.

"The longer you stay in one place, the greater the probability of seeing something unusual," Barlow said. "I think the animals also got accustomed to us, so we had a better chance to see typical behavior. But there were disadvantages, too. We were out of sight of the habitat almost all the time, so we had to operate on compass and really learn our environment, because if we ever got lost, that could be very dangerous."

Five men—four divers and an engineer—committed for two weeks to living quarters that essentially consisted of a pair of large, connecting steel tubes buried under almost sixty feet of water: to me, such an expedition is unimaginable, but then I get acute claustrophobia when the subway train stalls in the tunnel. I asked Barlow whether it bothered him.

"Some people can get a little twitchy," he said, "but actually I was very relaxed. In the evenings, we'd get together and have a can of beer and whatever the microwave oven would kick out for us. We slept on water beds, and I thought that was neat. I'd like to have the chance to do it again sometime." In the meanwhile, though, his schedule calls for scuba observations combined with lots of lab experiments.

In the complex of high-ceilinged rooms where he works at Berkeley, Professor Barlow and his students keep enough different kinds of fish to stock a tropical-fish store many times over. I was particularly intrigued by a large, circular tank shrouded by curtains, with a peephole at the front. It looked like an underwater arena, and it turned out that that was exactly what it was: it's where graduate student Terry Lim studies the way predator fish prey on schooling species. It seems most zoologists agree that schooling is a defensive mechanism, but nobody is sure exactly how it works. Terry has been observing—and will eventually videotape—attacks by *Haplochromis polystigma,* a beige-and-white African cichlid, on schooling zebra danios. Predation is very common, of course, in nature, but divers seldom see it happen because it's over so quickly.

Professor Barlow showed me, in a room crowded with aquariums, his Midas cichlids; some of them were huge, as tropical fish go. Barlow pointed out one male, obviously a favorite, that he caught back in 1965. At the age of about eleven, then, this animal was a gorgeous golden orange and almost a foot long. He may well live to be twenty and, since fish go on growing for as long as they live, attain a considerable size.

A few of the Midases in the room were gold, but most were black and white. Except for their shape, the two varieties looked so different that it was hard to believe they belonged to the same species. The question that interests Barlow at the moment is: what are the advantages and disadvantages of being gold? He has already learned that the golds dominate: in a fight with a black-and-white Midas of equal size, a gold has a 75 percent chance of winning.

"The Midas is an extremely intelligent and sensitive animal," Professor Barlow told me, "so sensitive that sometimes it becomes fear-struck. You have to know them to work with them—you have to be able to look at an animal during an experiment and recognize what's happening if, for some reason, it's not behaving normally."

Sensitive they may be, but they're also aggressive, which creates another kind of problem: each of the larger fish must have its own separate tank, or mayhem ensues. In nature, the Midas is quite gregarious and even forms loose schools, but in nature, of course, two fish can forestall a fight by retreating. The fiercest confrontations actually take place between mated pairs, who will sometimes battle over a breeding site, two against two, for an hour or more; they dart in to tear at one another's fins, or they lock jaws and hold on, twisting and turning.

To demonstrate cichlid fighting techniques, Professor Barlow leaned down and put his face up close against the glass of the tank that held the eleven-year-old male. He made a big round O with his mouth, protruding his lips, and the fish immediately swam over to him and, making the same kind of mouth, bounced it off the glass as if trying to bite him.

"I once had a lab assistant who may have been the only human ever to lose a mouth-fight to a cichlid," Barlow said, straightening up. "He was cleaning a tank one day and he leaned over the top and made a mouth; the fish leaped right out of the water and grabbed him by the lips!"

Cichlids, obviously, are interesting fish to have around. Not only do they fight and court, but they also raise and tend their young, rather than just abandoning their eggs to their fate, as many species do. This is just as well, since in nature, Midas fry lead a precarious existence. Barlow and Kenneth McKaye, one of his students, discovered that the half-life of a school of fry is about three days: in other words, of the one thousand to two thousand tiny fish that originally hatch, half are picked off within that time. And if the parents are evicted from the breeding site, as often happens, small, lurking predators generally polish off the whole school within the first five minutes.

Midas parents actually feed their fry by a method a little like nursing: when food is scarce, as a substitute, the tiny fish nip bits of mucus from the bodies of both father and mother. Midas parents are incredibly tolerant with their brood. Barlow showed me one female, then being nursed back to health in the sick-bay tank, who had had one fin stripped literally to the bones—which stuck up from her back spiky and bare—because her young were accidentally left in the tank with her for too long and they developed teeth that were all too effective. "When that begins to happen, the parents usually try to hide," Barlow said. "They'll go through all kinds of contortions, but they'll never, never hurt their young."

We settled down after that to the subject of communication. I knew that fish signal one another with visual displays and sometimes with sounds, but I was surprised when Professor Barlow explained that in his opinion, pheromones—chemical messages—may be as important to fish as they notoriously are to insects, though most fish probably have a smaller repertoire of such signals than an ant does, for example. Research on fish pheromones is still in the beginning stages, but we know already that some species use them for alarm signals, and that, for many, they seem to play an important part in mating. In the Midas cichlid, for example, the two sexes look so much alike that the only way Barlow can tell male from female is to probe for an extra little pore that the female has in the genital papilla; yet the fish themselves have no difficulty distinguishing a possible rival from a potential mate, presumably because they smell different.

Barlow believes that many cichlids also depend on chemical cues to recognize their own species, which is, of course, vital at breeding time. Furthermore, in lab experiments, Kenneth McKaye and Barlow demonstrated that Midas parents can recognize their fry by their smell and can even distinguish their own particular offspring from those of other Midas cichlids. Midas parents herd and protect their fry for six to eight weeks. They sometimes live in water so murky that visibility is almost zero, and even in clear water marauding catfish occasionally scatter the young at night. Under the circumstances, the parental ability to nose out the young is obviously important, and it's backstopped by the fact that the fry, in their turn, can use *their* sense of smell to locate their parents. Another of Barlow's students, Colin Barnett, has now shown that young Midases less than ten days old recognize and swim toward the odor of a mother Midas, though they apparently can't distinguish between their own mother and someone else's. Curiously, at that early age, they don't respond to the odor of a father, though perhaps that's because mothers provide the closest care at that time.

Fish also communicate with sounds, as I've said. Cichlids grunt at one another during aggressive encounters, and for some other species, mating calls have been identified; there are probably alarm signals as well. Some fish, Barlow told me, have vocabularies slightly more elaborate than others, but none have anything to compare to the sound repertoires of birds and mammals.

Sound and smell are at least familiar senses to humans, but there are also fish that make use of a communication channel peculiarly their own. About seven hundred species produce electric fields, and although a few of these fish, such as the electric eel, can generate enough of a jolt to kill, the rest are only weakly electric and use their fields to locate obstacles and other animals and to signal to one another. (Even the eel has a secondary, turned-down field for locating and signaling.)

A fish can use its electrical channel to "say" any number of things. For instance, a study of the banded knifefish, a South American species that looks a bit like a striped letter opener, indicated that one knifefish recognizes another member of its species by the pattern of its electrical discharge. A dominant male threatens a rival by creating brief breaks in the steady hum of his signal, or if he really means business, he may simply hum faster. Subordinate fish avoid trouble by shutting up altogether.

As for visual communication, there are in the pitch-black depths of the sea a number of species that signal to one another with flashing lights. Even on tropical reefs in shallower waters, there is a nocturnal species that winks out signals from a kind of lamp under its eyes; by day it hides in caves.

This is one kind of visible message, but there is another that's even more fascinating, for fish can do something that none of the other vertebrates can do: they can change colors rapidly and well. Though this is often done in the interests of camouflage, it can also be a social signal. For example, there's one kind of Hawaiian surgeon fish, the kala, that's basically black—except when it does battle, and then sometimes its whole body turns sky blue. Fish ready for combat often intensify their colors, while fading usually presages a retreat, and a badly frightened fish may literally turn pale. According to Barlow, the *Badis badis,* a small maroon fish from India, actually stages color fights: two fish will stop three or four body lengths apart and hang there motionless for perhaps ten seconds, turning on fighting colors as they eye one another. Finally, one will pale—indicating that it is giving up—and swim away, leaving the other the victor.

Some fish have as many as eight to twelve recognizably different color patterns, and some can switch patterns with startling speed. Barlow has seen trigger fish reverse from black stripes on white to white stripes on black within just one or two seconds. The black-and-white Midas cichlid, which when breeding has black vertical stripes down its sides and, in the center of each stripe, a black spot, can turn the stripes and the spots on and off independently. When it is moving about in a group, it generally wears just a few of the spots, with no stripes, but when it's ready to breed, the stripes become very dark and distinct. In Barlow's lab, one darkly barred male was living next door to a female who showed only spots. As I watched the male, his stripes visibly darkened and faded and then darkened again, the color ebbing and returning as softly and regularly as breathing.

A color change such as this—one that acts as a signal—is known as a *display.* In scientific terms, a display is a stereotyped behavior pattern that has, in the course of evolution, became specialized for conveying information. The curious thing is that even the most highly social of the vertebrate animals seem to have no more than thirty or forty separate displays. However, these can convey much more than thirty or forty simple messages, since the amount of information provided can be expanded in various ways. For example, though some signals are discrete—like a switch that can only be either on or off—others are graded: the fighting colors that fish assume not only signal a tendency to do battle but indicate at any given instant how strong that tendency is.

There is another kind of visual display, too, and fish use it in combination with color changes. During hostilities, in courtship, and while tending their young, they often make relatively stereotyped movements that convey a message, much as miming conveys a message to humans. A Midas cichlid threatens by facing the enemy, then depressing its dorsal fin and opening its gill covers wide; or it may charge but then break off well before it makes contact. A cichlid in a fighting mood will sometimes also present its side, raise its fins, and undulate its body slowly back and forth, ending with a rapid, emphatic swing of the tail, as if to strike its opponent. In general, an animal threatens by making itself look bigger and more dangerous, and this is true of most vertebrates—of fish and also of birds, wolves, humans, and so on. The more hostile the animal, the more it will augment its size, while an individual bent on avoiding a scrap is apt to adopt a low profile.

Like any other parents, fish that care for their young need a way to call them together, and, in general, they use a movment display. Cichlids simultaneously snap their heads to one side and flap their pelvic fins open in a motion that looks like an exaggeration of the first movements a fish makes when it intends to swim away; here, however, it's coupled with an abrupt halt. Presumably the signal originated as swimming away but has gradually become ritualized. The orange chromide, a species found in Sri Lanka, uses more specialized movements: a high-speed flickering of the pelvic fins. By squatting down to get a fry's-eye view of the parental belly (the fry are generally down below), Professor Barlow discovered that as the pelvic fins flick open, they uncover a crystal-white area on the fish's orange belly. "It's like a shutter flashing a white light on and off," he said.

Breeding time calls for specialized courtship displays. For the fish, they serve several purposes. First, they help synchronize male and female during spawning, so that they make their deposits of sperm and eggs in the same place at the same time. Secondly, because they differ from one species to the next, they provide another way for the fish to identify its own kind and so avoid interbreeding; and third, they help overcome the individual's understandable reluctance to let another fish approach closely—understandable since, for the most part, when one fish swims right up to another, its intentions are far from friendly.

In some cases, the displays used in courtship are the same as those used in aggression—Midas couples, for example, threaten each other while forming a pair. Professor Barlow believes that the female threatens the male because she has to stall him, to dissuade him from attacking her until he has had time to figure out that she's female, that she's ready to breed, and that she's the right female for him. Barlow also believes, by the way, that the immediate function of the threat display is to *reduce* the probability that the other animal will attack. Of course, if one or the other doesn't back down, bloodshed may indeed follow.

Paul Loiselle, another student of Barlow's, has been studying the mating system and life cycle of the California desert pupfish, and toward the end of my interview with Professor Barlow, I dropped in on Loiselle in his lab. What's interesting about the pupfish is that it leks, much as some grouse do: the males congregate on a common spawning ground and compete for the best breeding sites.

"It's really a kind of sexual cafeteria," Mr. Loiselle explained. "They take their positions and waggle away, and the females come along and look them over."

Once the females arrive and a male spots one, he swims out and displays in front of her with his fins flapping; then he arcs back into his territory. If she is interested, she comes closer, and he repeats the performance until he has lured her onto his home ground. Then they lie side by side for a moment and shiver while she expels one or two eggs for him to fertilize. Over the course of the breeding season, which lasts almost six months, she will visit quite a few males and lay five hundred to six hundred eggs in all.

Compared to the Midas cichlid, pupfish are tiny creatures. The largest individuals are rarely more than two and a half inches long. Loiselle had several tanks of them in his lab; the females were a muted brown and gold, the males a garish silvery blue with orange fins. Loiselle considers the pupfish an ideal laboratory animal because it's tough and adaptable, and it breeds like a machine, even in an aquarium.

"Also," he said, "pupfish are small, and when you're doing laboratory research, small is good. It's good because you don't need as much space to get normal behavior, and it's good because it's not as expensive to keep a lot of them around. Animals like the Midas cichlid eat enough in a year to support a graduate student, and I'm not exaggerating. I'm sure if George could redesign his cichlids, he'd make them just a couple of inches long."

Back in George Barlow's office, I returned briefly to the subject of visual displays because I wanted to know whether the same color changes always go with particular movement displays. Barlow explained that, on the contrary, the two systems operate independently to a degree and can, in fact, combine to create a complex, compound message. During hostilities, for example, an animal may continue to tail-beat, circle, bite, and ram while growing pale in preparation for beating a retreat. Some Midas cichlids can turn jet black, and Professor Barlow has seen mother fish in nature that wore black as long as they were back under the rocks and became boldly barred whenever they came to the entrance of their caves; yet in both places and color incarnations, they were going through the same motions—fanning their eggs, or protecting their fry, or threatening intruders, while the color message changed from cautious camouflage to bold warning stripes. He also mentioned that many cichlids, while they are parents, send two different color messages at the same time, wearing dark camouflage on their backs in shallow water to hide them from birds above and warning stripes on their sides to scare off small predators. Colors are often held longer than movement displays and seem to denote a more general state, while the movements make a kind of qualifying statement.

I also asked Barlow whether there is ever communication between fish that belong to different species. He explained that this hasn't been studied much, since researchers have mostly observed fish in aquariums where species weren't mixed. However, from diving observations on coral reefs, Barlow has learned that interspecific communication is actually common. "There's a lot of signaling back and forth," he told me, "and it's not just predators going after prey but lots of interaction, competition, and aggression." It seems that, although some signals are species-specific—particularly color displays—others, such as behavior patterns, are similar enough from one breed of fish to the next to serve as a kind of Esperanto.

How smart are fish? This would seem a relevant question, since presumably intelligence affects what they have to communicate about. Looking only at brain structure, some psychologists have found no differences that would suggest that the alligator, for example, is more intelligent than the shark or the salmon.

However, it seems that it's difficult to speak in a meaningful way about comparative intelligence in animals, for each species has evolved to fit into a particular environment and has developed the skills that will help it survive there. Maze-running may be a useful skill for a rat, but it has little relevance for a shark cruising the open sea. To take another example, goldfish learn quite rapidly that when a warning buzzer sounds, they must swim from one compartment of an aquarium to another to avoid receiving a mild electric shock. Siamese fighting fish learn this much more slowly. However, this doesn't mean the goldfish is brainier; it is simply the kind of fish that reacts to sudden stimulation by fleeing for its life. The Siamese, on the other hand, is a bold little predator who is more inclined to stay and fight.

Comparisons aside, it's obvious that fish have a memory capacity and can learn. Professor Barlow believes that many of them can also recognize and react to other fish as individuals. "We wouldn't anticipate individual recognition in schooling fish—in anchovies or sardines, for example," he told me, "but we'd predict it in cichlids, and six to ten neon tetras in a home aquarium probably learn to distinguish one another."

It also seems quite likely that the fish in an aquarium can learn to recognize the human who feeds them. Paul Loiselle showed me some African cichlids in his lab, and the minute he approached their tank, they came swimming frantically to the top at the end where he was standing. Remarking that they obviously expected to be fed, he told me that when a stranger approaches, they generally hover instead at the middle level and peer out through the glass.

A home tank can be like a window on an alien world: inside, small creatures fight and mate and raise their young before your eyes. There can be a lot to see if you know what to look for—and if you stock your tank with an interesting cast of characters. That's important, and it's the main reason my own tank was so dull. In the beginning, I selected fish for color, as if I were putting together a bouquet of flowers or arranging an underwater landscape. Beyond that, I insisted on peaceable species suitable to a community tank, because my children winced at bloodshed. Small wonder, then, that nothing ever seemed to happen in there.

Cichlids are probably the most interesting fish to keep, but in order to live peaceably together, they generally need a lot more space than the average aquarium provides. However, some of the easy-to-keep dwarf cichlids can be safely added to a community tank, and that's what Professor Barlow recommended. Provide them with a cavelike place where they can lay their eggs, and soft acidic water, and they will generally court and mate, and the female will care for the young. "She puts on quite a show," Barlow said. "She threatens the other fish and drives them away, calls to her young, and does all sorts of neat things, including some very graphic and dramatic fin and color displays." Though dwarf cichlids are small, one male and one female are all a ten- or fifteen-gallon tank will peaceably hold.

Given a tank stocked with this or other species, most aquarists soon discover that fish are not as predictable in their behavior as they may have been led to expect. A well-matched pair who should be ripe for breeding may inexplicably show no interest in one another. An individual who ought to be quite timid, going by the reputation of his species, turns out to be the bully of the tank. Male guppies can ignore their females and make a nuisance of themselves courting lemon tetras. The moral is, I suppose, that it's always risky to generalize. There's very little one can say about fish behavior that would actually be true of all of the twenty thousand known species of fish. More important, even within a species, there are differences between individuals—just as there are between humans.

9.

Army Ants: A Report from the Front Lines

A solitary ant, afield, cannot be considered to have much of anything on his mind; indeed, with only a few neurons strung together by fibers, he can't be imagined to have a mind at all, much less a thought. He is more like a ganglion on legs.

Ants are so much like human beings as to be an embarrassment. They farm fungi, raise aphids as livestock, launch armies into war, use chemical sprays to alarm and confuse enemies, capture slaves. . . . They exchange information ceaselessly. They do everything but watch television.

—THE LIVES OF A CELL,
by Lewis Thomas

I was lost in the American Museum of Natural History, and it wasn't exactly a first for me. I'd often used the museum's library, and I regularly got lost on the way there until I memorized the route: elevator to the fourth floor, then through Late Dinosaurs, Early Mammals, and Late Mammals to the Hall of Earth History and turn right.

But getting lost this time was a little different, since I was venturing now into the nonpublic areas of the museum in search of the Department of Animal Behavior. Following instructions, I had already taken an elevator to the right floor in the wrong wing, and I was currently trying again aboard a different elevator, this one almost the size of a freight car. As he opened the door to let me off, the sympathetic operator agreed that the Department of Animal Behavior *was* straight ahead all right, if I was sure this was the particular section of it I wanted.

But the moment I started down the corridor, I felt sure I was in the right place, for I could hear the cooing of ring doves through a partly open door. In addition, projecting above each office was a shingle with a name on it, and from my reading, some of the names were as familiar by now as the *Tyrannosaurus rex* I had passed so often in Late Dinosaurs. Sure enough, at the end of the corridor was the shingle of Howard Topoff, the man I had come to see.

Dr. Topoff turned out to be neatly bearded and youngish, with just enough gray in his dark hair to look rather distinguished. His particular animal is the army ant, a creature that has fascinated generations of scientists. In the world today, there are approximately 270 different species of army ants, and they all have two things in common: they carry out attacks in vast, well-organized columns, and they regularly break camp and emigrate. Most species are essentially blind, and most live out their lives underground; however, a few operate on the surface of the soil, and of course these have been the most thoroughly studied.

Sheer numbers make the army-ant colony an impressive phenomenon. Depending on the species, a small colony might contain ten thousand individuals; a large one, as many as thirty million. In jungles in tropical climates, the ants begin to pour forth from their bivouacs at daybreak, in vast columns that fork and then fork again. As they creep through the underbrush, accompanied by a soft, rather sinister rustling, they look like a great, sluggish, red-brown river. Soon there are workers traveling along the column in both directions, as those in the vanguard head back to the nest, carrying torn bits of insects or wasp larvae or other booty. As the day wears on, one of the raiding parties locates a new nest site, and the raid turns into an emigration; eventually the queen emerges from the old bivouac with her guards, and workers carrying the curly white larvae also head out along the trail to the new nest. For two or three weeks, this is the daily routine; then suddenly the emigrations stop, the ants settle down, and though raiding still goes on, it's smaller in scale and less vigorous; this is called the statary phase of the cycle. A few weeks later, the wheel has turned full circle and the daily migrations begin again.

In the popular imagination, army ants surge through the jungle like a tidal wave, consuming every living creature, animal or human, in their path. This, it turns out, is pure mythology, since the army ants' diet runs almost entirely to insects, with the occasional small lizard or snake thrown in. Dr. Topoff explained ruefully, "If they were pests, it would be easier to get money to study them." Nor are they particularly intelligent, as ants go—on the contrary, their behavior is more uniform and stereotyped than most species, and as individuals, they show very little independence. Nevertheless, scientists have long been intrigued by their remarkable life cycle and by the question of what controls it.

The particular species of army ant that Howard Topoff has been studying in recent years is *Neivamyrmex nigrescens*, an insect native to southern Arizona, where the museum's Southwest Research Station is located. With only ten thousand to fifty thousand workers to a colony, *N. nigrescens* is almost a scale-model army ant, much easier to study than some of the tropical species with their swarming millions. However, there are some drawbacks: *N. nigrescens* is nocturnal in its habits and nests underground. In the field, Dr. Topoff and his colleagues must don miners' headlamps to follow the nightly raiding columns, mapping routes, estimating booty, and so on by flashlight. Eventually the scientists move back into the laboratory to observe the behavior of captured ants and test out theories.

Until last summer, however, army ants had always proved extremely difficult to maintain in a lab. Dr. Topoff explained, "You can't just bring them in and put them in a nest the way you can carpenter ants, for example. We've tried that and their social organization falls apart. The queen stops laying eggs, they eat their own brood, and you get nothing that resembles army-ant behavior. Nobody has ever succeeded in getting them through a full nomadic and statary cycle except us—maybe. We did it once last summer, and next year we'll be trying again, to see if we can get it down pat.

"The problem is that here you have tens of thousands of predatory insects that feed on the brood of other insects, and it's difficult to get enough brood, since they eat several thousand pieces of booty every single day. So we brought them into the lab right there at the research station, where we could more easily go out and collect fresh food. We also set them up in a very species-typical way. They have to have nests, to be able to raid around the lab through plastic tubes—otherwise they fall apart."

As it happened, there was, at the time I talked to Dr. Topoff, an experimental ants' nest complete with raiding tubes set up in a lab across the hall, and he took me over to see it. Ellen Kwait, a graduate student, was doing research on slave-making ants, and the room was filled with an amazing network of clear plastic bins connected by clear plastic pipes. The bins sat on tables at about waist level, and the pipes formed elevated runways between them. In the far corner was a small curtained-off area, like a photographer's makeshift darkroom: the home of the slave-making ants. Behind each plastic bin was something that looked like a shoebox, covered with a piece of heavy fabric. There were eleven of these in all, and they were the homes of ants that slave-makers take as slaves—stealing the young in their cocoons and carrying them off to become worker-servants in the nests of their captors.

The bins and their connecting pipes were floored with sandy soil, and a few ants could be seen in each, but for the most part the apparatus was deserted. Late in the afternoon, the slave-makers would emerge from their darkened nest behind the curtains and set off down the plastic pipes to raid the shoebox colonies.

In all, there were only a few hundred slave-makers—a viable size for a lab colony with that species—and Ms. Kwait had marked each with three dots in different colors so that she could identify individuals. She wanted to find out how slave-making ants decide which colony to raid. Do they, for example, send out scouts who inspect the nests available and then come back to lead the raid? She has found that in the lab, with a fixed number of colonies to choose from, the ants soon learn where all of them are and no longer need a scout. Now it's time to go back out into the field to see if this holds true there, as well.

As we returned to his office, I asked Dr. Topoff if any of his laboratory ants ever turned up missing. He said drily, "You know, most of the people in this department work with fish, rats, and pigeons, and they have always boasted that they've never lost an experimental animal. Well, I lost thirty thousand of them in one day."

The incident occurred back in the late 1960s, the first time Topoff ever tried to bring army ants back to New York from Arizona. He had wooden boxes and runways ready and waiting at the museum and, arriving at the airport late at night, rushed the ants into the city to deposit them in their new home. However, the connection between one box and its runway was uncorked, and the next morning when Dr. Topoff arrived at the lab, there wasn't an army ant to be seen.

"When I told the head of the department what had happened," he recalled, "he was panic-stricken. He was sure the director of the museum would be sitting in his office one day and an army of ants would come marching out of the electrical conduits. We spent three or four nights going over every nook and cranny in the department and never saw a single ant. We had a few hundred left that we'd isolated, I don't remember why, and we even tried putting them down, figuring if there was an odor trail they might lead us to the rest of the ants, but they didn't know where to go, either. That was eight years ago, and for all I know, that colony is still in the basement somewhere, raiding cockroaches."

We turned then to the subject of pheromones, since that's the primary channel of communication for ants. I had already learned that ants signal with chemicals in a number of different situations. There are, for example, alarm pheromones that can send workers scurrying for cover or provoke them into an attack. Most of the time, the insects release these substances in such minute quantities that humans can't smell them. However, Dr. Topoff explained that when an army-ant nest in the field is disturbed—so that there are perhaps half a million individuals simultaneously giving off alarm pheromones—there's no missing the odor, which in some species is distinctly lemony.

Slave-making ants such as the ones Ellen Kwait is studying actually produce something that has been called a propaganda substance, a pheromone very similar to the alarm signal of the slave species they raid. The slave-makers themselves merely find it exciting and an attractant, but it creates total panic in their victims.

Every colony of army ants also has its own colony odor, generated largely by the queen. She is constantly licked and groomed by the workers, who also groom one another and the brood, and so her pheromones are passed around. If an ant fresh from another colony enters the nest, it's apt to be ejected or killed because it doesn't smell as if it belongs.

There are almost certainly sex pheromones, as well, by which virgin ant queens attract males, but the most remarkable chemical signal in the ant repertoire is the trail pheromone. Army ants in particular, since most are pretty nearly blind, are heavily dependent on it, and when they're on the march, they swing along like bloodhounds, sweeping the ground in front of them with their antennae, because that's where their "noses" are. An ant at the front of the column, stepping out onto virgin ground on which no trail scent has yet been deposited, reacts as if shocked. She hesitates (all worker ants are females), then crouches and crawls forward, tapping the soil nervously with her antennae. As she goes, she rubs her abdomen against the ground, releasing the trail substance. She proceeds for only a few centimeters, then retreats, and another ant takes her place.

Scouts regularly leave the raiding column, and when one locates food, she lays an odor trail as she returns to the column. Rejoining the raiding party, she runs up and down, touching antennae with other ants. It is apparently this physical contact that recruits workers and sends them out along her trail, for if she is captured after she has laid the trail but before she has done her antenna-touching routine, very few ants are recruited.

When I asked Dr. Topoff whether any of the army-ant pheromones have been chemically analyzed, he shook his head. It seems a chemist at the University of Chicago has been working on the trail pheromone, but it's a slow process. Army ants lay a trail by depositing their feces, which may or may not contain one or more pheromones produced by glands in the gut. On the assumption that a chemical signal *is* included, Topoff collects the fecal material and sends it to the chemist, who distills it until he has a few drops of extract that he can assay with sensitive instruments. Eventually he returns to Topoff fractions of the original material, to be tested for trail-following with live ants. Once an extract has been found that seems to work, the next step will be to try to identify its chemical components. "It's a very slow and tedious procedure," Dr. Topoff said. "I'm constantly amazed that chemists can work with quantities that small."

Edward O. Wilson, Harvard professor and author of the book

Sociobiology, has suggested that as few as ten pheromones, used singly and in simple combinations, may be all it takes to organize an ant colony, and Dr. Topoff is inclined to agree. When I remarked that this seemed amazing, since ant behavior can be so complex, he explained, "The complexity is in the interaction. The behavior of the individual ant is very, very simple."

And actually, reading over papers published by Topoff and by the late T. C. Schneirla, who was once curator of the Department of Animal Behavior at the museum, I began to feel that an army-ant colony is most like a vast Rube Goldberg contraption, with thousands of moving parts. Take the forces that shift the cycle from the statary to the nomadic phase, for example. It isn't hunger that gets the colony moving, nor chemical commands from the queen; it's the development of the brood that triggers the change. When the time comes for the fully formed young ants—called callows—to emerge from their pupal cases, there is great excitement in the nest. Adult ants gather around and eagerly pull off the casings and then set to work lapping up secretions from the callows' bodies. These secretions apparently contain pheromones that kick off the nomadic phase, for as soon as most of the callows have emerged, the adult ants set out on a massive raid.

As the callows mature, their colleagues no longer find them as stimulating, but by then, the most recent brood of eggs has developed into thousands of squirming larvae, and to the adult ants, these are just as exciting. They respond to them by feeding and grooming them—or by running off to join a raid. However, once the larvae begin to pupate—to wrap themselves in their protective cases—the excitement in the nest decreases sharply and the statary phase begins. Dr. Schneirla suggested that because the pupae don't need to be fed, there is more food for the queen, who also gets more attention from workers freed from nursery chores; this state of affairs may well be what triggers egg-laying in her, and the beginning of the next generation.

The queen and her colony are indispensable to one another—neither can survive without the other. Deprived of its queen, a colony deteriorates steadily and soon becomes extinct; while the queen, if she is removed from the nest, lives only a few days. Apparently she's unable to get along without the feeding and almost incessant grooming she receives from the workers, and perhaps she also needs their secretions.

This kind of interdependence is common, at least to some degree, in all species of ants, and according to Dr. Topoff, it's the reason commercially sold ant farms are so disappointing. Usually they come supplied with two dozen or so little red harvester ants, where a normal colony would have hundreds if not thousand s of workers plus a queen. With no queen, no brood, and just a thimbleful of workers, it's not a real colony, and so you see very little of the typical ant social organization; the animals also tend to die off rather quickly. "I would advise anyone who wants an ant farm to go outside and dig up a colony," Dr. Topoff said. "You turn over rocks until you find one, and then all you need are one or two scoopfuls, though you have to be sure you get the queen. You also have to have some idea of what they eat."

Ants in an ant farm are one thing, however, and ants loose in the pantry are another. I remember I once tried—unsuccessfully—to coexist with ants in a beach house. It wasn't squeamishness that made me hesitate to buy ant traps, as my housemate suggested. I had some idea that I could watch them and see how they lived. But the morning I poured out the children's cereal and it came out half Cheerios and half ants, I gave in. I realized that not only were they competing with us for the groceries, but they were winning.

There is, as Dr. Topoff pointed out, great ambivalence among humans when it comes to insects. They are in many ways our most formidable animal competition—supposedly they consume, destroy, or render inedible about half of what we produce. Except for the honey bee, which is indispensable, we have largely been concerned with how to wipe them out.

And, in fact, in a major new thrust of research today, scientists are using what they've learned about insect pheromones in an effort to suppress pests. For the most part, they're working with sex attractants and taking advantage of the truly remarkable signal systems that insects have evolved.

Ants, as we've seen, use a whole battery of pheromones, but for economy and reliable performance, moths seem to be the real prodigies of smell, and of course some of them *are* pests. The male moth has a sense of smell so acute that a minute amount of the chemical aphrodisiac produced by the female will fire his olfactory nerve. Thanks to this sensitivity, she can squeeze a few drops of scent from glands in her abdomen and with it call downwind, reaching males literally miles away. Receiving the signal, they simply fly upwind until they find her. To a male moth, however, the pheromone the female produces, to all intents and purposes, *is* the female, for if a scientist snips off her scent glands—which leaves her looking and acting exactly as usual—the male completely ignores her and tries to mate with the amputated scent glands instead.

Obviously, if we can learn to synthesize insect sex pheromones, they might have tremendous potential as an alternative to pesticides, both because a little would go a long way and because they could be used to tap into that blind, automated, sexual response. Researchers are experimenting right now with several different techniques. In some cases, they're simply using a sex pheromone to bait small traps, so that the pest level in an area can be monitored and insecticides applied only when a real problem is about to develop. In other studies, an orchard or field has been permeated with the sex scent of the female, so that it smells to males as if females are everywhere; thrown into confusion, most fail to mate and reproduce. Sex pheromones have also been used successfully in forests to entice western pine beetles to a sticky death on panels coated with goo. So far, the work has been strictly experimental, and there are still significant practical problems to be overcome. Identifying and synthesizing a pheromone is, as Dr. Topoff pointed out, a difficult and time-consuming process, and since the chemical ultimately produced can only be used against a single species, developing one isn't as profitable as concocting an ordinary insecticide. However, because synethetic pheromones are unlikely to poison the countryside, and because they *do* affect only the target insect and won't wipe out whole populations of innocent bystanders (such as bald eagles and bullfrogs), they're an environmentalist's dream.

From the point of view of those interested in animal communication, the new biochemical research also has another significance. In effect, we have learned how to lie to some insects, using their own pheromones. In the long and contentious relationship between humans and insects, that has to be a milestone.

10. The Dance Language of the Bees

A moment's wild scramble and I was through the bedroom, dashing over the furniture, had gained the bathroom and slammed the door. Rushing to the window, I slammed that down. Outside the door I heard the angry buzz and even the sinister little taps of the bees flinging themselves in murderous frenzy on the panels. A moment later I saw a couple come crawling from under the door. I stamped on them and felt an unpleasant glee as their bodies crunched on the tile floor. They were deadly as flying snakes, but my heel could still bruise their heads once they were forced to crawl.

—A TASTE FOR HONEY,
by H. F. Heard

How doth the little busy bee
Improve each shining hour . . .

—AGAINST IDLENESS AND MISCHIEF,
by Isaac Watts

Through an observation window, the video camera had caught half a dozen citizens of the hive in a tight close-up. Wriggling, furry, golden bodies filled the screen in what was surely a bee's-eye view of the inside of a hive.

In the center of the group, a dancing bee was going through her routine, evidently just back from scouting up some food. First she crawled in a half-circle to her right; then suddenly she shot ahead, wagging her whole body rapidly from side to side. Just as suddenly, she stopped wagging and crawled patiently in a half-circle to the left, only to repeat the wagging run again. I knew that the direction of that run indicated the direction of the food she'd found, and that the harsh, staticlike sound she made was a clue to its distance. As she reversed herself and began a third run, the videotape stopped, freezing her in her tracks.

"You see," said Professor James Gould, "the angle of her run is quite different this time than it was before." Mounted over the screen was a large, transparent, plastic disk with rows of parallel lines etched in white on it. Professor Gould spun the disk until the lines fit over the bee's body. It was clear that she was headed up the screen in a north-northeasterly direction. Then he ran the tape backward and did an instant replay of the previous waggling run. When he fixed her again between the parallel lines, she was moving almost directly north.

"You have to do a lot of averaging to figure out what direction the food is really in," Professor Gould explained. "If you're von Frisch, you sit and average for two hours. If you're a bee, you average for six cycles of one dance." Von Frisch is, of course, Karl von Frisch, the famous German scientist who originally discovered the dance language of the bees. It seems he once had his students hide the bee-feeding station so that his only clue to its location would be the dances of scout bees back at the hive, and it took him two hours to feel confident that he could correctly interpret their signals, though he did eventually get them right.

The videotape was running again, the dancer was dancing, and the audience milling about. I realized that many of the bees wore white number tags on their backs, like track stars, making it possible to follow the behavior of individuals.

"You see," Professor Gould said, "number 12 there is trying to watch the dance, but she just got cut off by number 31. Thirty-one is following it now, and she's not doing too badly. There's the stop signal—that little beep. It means, 'Stop and give me a sample of food or I'll kill you.' And they will, too, if she ignores the signal for very long.

"You can see that a lot of the audience is off to the side and can't really be getting much impression of the dance. Those who do manage to get close keep getting in each other's way. Then, as I said, they have to average the direction signal; the distance signal, the buzzing, varies from one run to the next, too, so that also has to be averaged. What's impressive is the chaos. It's amazing that they're able to get any information at all from the dance."

In recent years, the dance language has been the subject of a lively controversy, with some American scientists insisting that it serves no communicative purpose at all. James Gould, now a professor at Princeton University, has the distinction of having performed the definitive experiment, the one that proved that bees really do use the information coded in the dance. A lean, towering, dark-haired young man, Gould was going to be a molecular biologist until he tried a bee experiment one summer as a lark. Bees are now his animal, despite the fact that he recently discovered that he's allergic to bee stings.

"I'm having shots," he explained. "They're supposed to work for 90 percent of those who try them. And, you know, Frisch is allergic to bee stings, too."

Professor Gould's top-priority project right now is the construction of a model bee that could be sent into the hive to communicate with the regular bees. For twenty years, scientists have been trying to make such a model, but in most of the early experiments, it was immediately executed as a spy. Finally a German researcher, Harald Esch, now at Notre Dame, figured out how to make one that would at least not antagonize: he coated a dead bee with wax and left it in the hive until it had absorbed the hive's odor. (Dead bees apparently smell dead and, uncoated, are carted out of the hive in short order.) His model wasn't attacked, but when he attached it to a mechanical device that made it waggle, the bees ignored it. A small, wax-coated microphone playing the dance sound aroused more interest, but it soon overheated and drove the audience away.

It's actually quite difficult to fake the buzzing noise that dancing bees make by popping their wing muscles, and the fact that the audience expects the dancer to provide food samples on demand is an additional problem. Professor Gould once tried to get around this by attaching a supply of sugar water, dispensable through a tiny plastic tube, to a wax-coated chunk of wood, but he found that even after he stuck on wire antennae, the bees couldn't tell the back from the front of his model and kept begging from the wrong end.

Gould's current approach is to use a living bee that has been paralyzed. He attaches electrodes to the insect's wing muscles and uses very weak electric current to pop them and make the dance sound, and he dispenses food samples through plastic tubing that the bee wears. The model is put into the hive through a trap door and remains connected by its tube and wires to the experimenters outside.

"One of the nicer things about this is that my models usually recover," Gould said. "After we remove the apparatus, they crawl out of the little straps and back into the hive."

So far, the audience does beg from the right end of Gould's model, and bees have accepted samples from it, but more work still needs to be done. However, Gould believes that within the next few years, he will succeed in creating a model acceptable to the bees. Then he will be able to study the dance language in much finer detail. He can vary the rate of waggling, for example, and see whether the recruit bees respond by heading off in a slightly different direction, perhaps, or traveling a different distance.

The dance language was originally discovered, as I've said, by Karl von Frisch. In the 1940s, he concluded that the dances that forager bees perform when they return to the hive spell out, in a simple code, the location of the food they've found. If the feast is nearby, the dancer simply whirls in tight circles. If it's farther away, she switches to the waggle dance, and the direction of her waggling run symbolizes the direction of the food. Let's say, for example, that the direct route from the hive to a rose bed would lie about twenty degrees to the right of an imaginary line drawn over the ground in the direction of the sun. Back in the hive, dancing in the dark on the vertical comb, the scout bee substitutes the pull of gravity for the sight of the sun and makes her wagging run heading up the comb at an angle twenty degrees to the right of the vertical. If, instead, the route lay directly toward the sun, she would make her run straight up; if away from the sun, straight down, and so on. She also provides a number of different clues to distance—for example, the farther away the food is, the more time she will spend buzzing during each cycle. Even the quality of the find is reported on—by the liveliness of the dance. Not only is the dance language clearly symbolic communication, but it also does what only human language is supposed to be able to do: it describes something not present, something distant in both time and space.

It would seem improbable enough if we discovered that a squirrel monkey, for example, had a way to say, "North by northeast, three hundred meters, a four-star find." It's even harder to believe that an insect—a tiny, six-legged creature—can send such a message. Not surprisingly, in the 1960s, some American scientists, including Professor Adrian Wenner, challenged the dance-language theory. Wenner insisted that recruit bees locate food in the field by odor alone, using both the smell of the site itself, which is brought back to the hive on the body of the scout bee, and odors that the scout bee actually uses to mark the site. He suggested that recruited bees simply drop downwind from the hive until they pick up the scents they're looking for. From previous flights, they know the olfactory landscape, and that makes the task even easier: they may simply remember where to find nasturtiums or the experimenter's peppermint-flavored sugar water. When Wenner and his colleagues set out feeding stations that all smelled the same, sure enough, *their* bees showed no preference for the one whose location had been danced but turned up at all the stations.

Certainly, forager bees do mark a food find both with a pheromone released from an abdominal gland and with so-called footprint substance, which comes from glands in the bottom segments of their legs. In addition, each hive has its own distinctive odor (just as ant colonies do), shared by all the inmates, and this, too, probably clings to the feeding station or flower.

It's also true that other insects behave much as the bee does and apparently communicate nothing at all. Some ants, beetles, and spiders will transpose a direction with regard to the sun into a direction with regard to gravity, though they convey no message by doing it. Stingless bees do a buzzing run, and the sound bursts correlate with the distance of the food find. Yet it's obvious that recruits don't use that information, since on her flight back to the hive, the forager bee touches down at intervals to lay a scent trail;

then, to be doubly sure, she actually leads her colleagues to the banquet in person.

There is probably a physiological explanation for the buzzing run—it may, for example, reflect the amount of effort it took to fly to the site, expressed as nervous energy. It's possible that there's also a physiological explanation for the ability to transpose between light and gravity. All the same, it seems likely that in the case of the honey bee, evolution leapt on these correlations and used them to create a system of communication.

Some American scientists never really took Wenner's criticisms seriously, but Gould said, "I *thought* I knew he was wrong, but there was no conclusive proof, and that made me a little nervous that someday someone might find out he was right." On the one hand, there was nothing bees had ever been observed to do that couldn't be explained by the odor theory. On the other, even if they located food by odors, that didn't necessarily mean they didn't also have a dance language.

Gould reasoned that if he could get a forager to lie about where the food lay and if the bees, believing her, flew off to the site she described in her dance rather than to the real site, that would be conclusive. To get bees to lie, he took advantage of several facts. First, if they can see the sun, bees don't bother to reorient their dances to gravity; instead, if the food lies 90 degrees to the right of the direction of the sun, dancing on the hive, they point their waggling runs 90 degrees to the right of the visible sun. Second, when an observation hive with glass windows in its sides is placed inside a dark shed and a bright light is aimed at it, the bees react as if that light were the sun. Third, by painting over a bee's ocelli—the three simple eyes between the big compound eyes—an experimenter can make the insect much less sensitive to light, though it can still see well enough to get around.

By trial and error, then, Gould found an amount of light just intense enough so that foragers who had had their ocelli painted danced as if they were in a dark hive and oriented to gravity, while their audience oriented to the light. From there on, the experiment worked this way: assume that the food lay in the direction of the sun and that the light was aimed at the hive, not from directly overhead but from a point 90 degrees to the left of the vertical. The painted forager, cued to gravity, makes her runs directly up the comb, meaning: fly toward the sun. But the watching recruits can see perfectly well that the "sun" is 90 degrees to the left of the vertical, which means that as far as they are concerned, she is saying: fly 90 degrees to the right of the sun. Since this *was* the direction they flew in when they left the hive, that proved they used the information coded in the dance. By moving the light around, Gould was able to send the recruits off in any direction he pleased.

Gould believes that when there is a single, abundant crop available, honey bees may at first use the dance language to alert recruits to its existence and location. After that, as more and more foragers return from the field and the odor of the crop accumulates in the hive, the need for dance recruitment may disappear while the bees find their way by odor cues alone. Since in the tropical forests in which the honey bee and its language evolved, food was most likely to occur in distant, isolated patches, there the dance language would have constituted a real advantage.

Gould's more recent research has concentrated on following the behavior of individual bees. He starts by numbering the flying workers in a hive, catching them himself as they appear at the hive entrance, while two assistants wait to tag them with numbers. His capture technique is simple but effective: he pops a small plastic bag over the bee's head, and when she walks up the side of the bag, closes it with a wire twist and immediately drops it into an ice chest. Chilling, he said, wipes out the insect's short-term memory, erasing its recollection of the ten minutes before the temperature dropped. If you wait ten minutes and then begin chilling, the difference is quite dramatic, since the bee wakes up mad as a hornet.

As for the tagging, Gould explained that he and his crew can comfortably number about a thousand bees a day "if we don't work too hard and take frequent breaks. If you don't take breaks, you start getting sloppy. You get glue on the wings, or you're careless catching the bees, so you get stung. Or you try thawing too many bees at once. You see, you have to get them back into the cold before they wake up and start walking around. So you dump out thirty bees and then two people sit there, each going dot with the glue and dot with a number. If you get a little too sure of yourself, which comes of being tired, things can get out of hand and there are bees flying around."

The next step is to train a couple of bees out to a feeding station, which is done by greeting them at the hive entrance with a solution of sucrose and then gradually luring them farther and farther away. Professor Gould starts with a diluted solution, so that the bees won't work up enough enthusiasm to dance about it back in the hive. Once he's ready to begin the experiment, he makes the solution stronger, and the forager bees start to dance. Since Gould wants to know which bees watch the dance and what they do about it, he videotapes the action at the hive; he also posts assistants at the feeding stations out in the field to capture recruit bees as they arrive. Eventually, back in his office at Princeton, he compares the videotapes, which are computer analyzed, with the field reports. He has found that invariably the bees that do show up at the feeding station did watch the dance, but on the other hand, few of those who watch turn up at the stations. Many seem to attend the dance the way some people watch television—just to pass the time—and they don't even leave the hive to search for the food. But of those who do leave the hive, more than half find the food source within about six minutes.

It's clear from the data that the bees do average the dance information. Watching and averaging with his own equipment, Gould has concluded that they would have to watch at least four cycles of a dance to get the direction and distance right, and the magic number actually seems to be six or perhaps six and a half cycles. A bee who watches more than six cycles does no better than one who quits after six, though a bee who watches four does significantly worse.

Another thing Gould has discovered is that bees are both idiosyncratic and predictable, for once they've been numbered, he explained, "you begin to recognize your old friend white 84." One bee, for example is always brief and businesslike; another feeds at leisure and then scouts around the outside of the dish for drips. Standing by a feeding station, Gould could often predict with his eyes closed which insect was coming in: this was the one who always circled twice, or the one with the very throaty buzz. New recruits were easy to spot—they came in low to the ground and, because they were hovering and hesitant, the tone of their buzzing was slightly different.

The favorite bee in the summer of 1976 was the one dubbed the bicentennial bee, because she was number 76 and because red, white, and blue stripes had been painted on her abdomen—

indicating that at one time or another, she had visited the feeding stations coded red, white, and blue. Covered with paint, a real veteran, she would fly first to the station at which she'd started her experimental career, circle it, then head for the last station to feed; back in the hive, her dances always pointed directly to the last station.

Just as the experimenters could tell the bees apart, the bees soon learned to tell the humans apart, apparently by smell. If an observer who had been watching at a feeding station walked off into the bushes to relieve himself, very often his bees, not finding him at the station, would come to him in the bushes. Foragers that Professor Gould had been training would come looking for him first thing in the morning, and if someone else came out onto the field, they would fly to him, take one sniff, and then leave.

Bees only live about six weeks, and Gould found that he could actually see them getting older. Suddenly they were not quite so good at foraging. During the last few days, they were apt to be blown off the feeder by the wind, "and you'd know," Gould said, "that soon an old friend was no longer going to be there. Sure enough, the day would come and old white 84 just wouldn't appear."

In the future, Professor Gould plans to look into the matter of bee bias. It seems that some bees consistently dance that the food is farther away than it actually is, or nearer, and when they interpret a dance, they consistently err in the same direction. It appears that each insect is born with a formula, probably genetically coded, that translates distance into the waggling run and that operates both for dancing and for interpreting.

Gould also wants to track recruits flying in the field. One of the mysteries about bees is why it takes them so long after they leave the hive to find the feeding station. Tracking a recruit visually is almost impossible, so Gould will trim the wings of particular bees in such a way that their buzzing is distinctively higher in tone, and then he will stud his experimental field with microphones and use a computer to plot each insect's course.

And, of course, there is also his work on the model bee to keep him busy.

In a remarkable book called *The Question of Animal Awareness*, Professor Donald R. Griffin of Rockefeller University argues that the mental experiences of animals and humans *may* be much more alike than most humans are ready to believe; that animals may, after all, work from mental images of past, present, and future events and have an awareness of the world that is at least on a continuum with human awareness. Professor Griffin doesn't insist that this is so; he only suggests that these are possibilities that should be investigated. In the past, scientists have avoided such questions, partly because they assumed that there was no way to know what an animal's mental experiences are like.

In suggesting that we may have underestimated animals, Professor Griffin often cites what is known about bees—describing, for example, the way consensus is arrived at during swarming. Soon after leaving the hive, the bees settle on a tree branch or some such place, and from there, scouts go out to look for a suitable location for a new hive. Returning, they dance their directions on the side of the swarm. After delivering her report, a scout may attend the dance of another bee and fly off to inspect the site described. If she becomes a convert, she may then return to the swarm to dance the new location. Griffin writes that "Only after many hours of such exchanges of information, involving dozens of bees, and only when the dances of virtually all the scouts indicate the same hive site, does the swarm as a whole fly off to it. . . . This consensus results from communicative interactions between individual bees which alternately 'speak' and 'listen.' But this impressive analogy to human linguistic exchanges is not even mentioned by most behavioral scientists. . . . "

Griffin also notes that behaviorists often define *thinking* as "covert verbal behavior"—in other words, we think by silently talking to ourselves. In a way, bees do this, too. If you flash a light on a hive in the middle of the night, the rare, occasional bee will begin to dance. Its dance will be brief and inaccurate, but it will indicate a food source visited the day before. This suggests, Griffin said, that the memory of the food location and the motivation to communicate about it are both still present in a latent, covert state. So if thinking is covert symbolic behavior, does the bee think? Or is this just playing games with definitions?

Because so many of Griffin's arguments cite bees, I was eager to find out how James Gould felt about them. Is the dance language really a language, I asked him. And what about the mental experiences of bees? Are bees conscious, self-aware? Or are they furry little robots whose genes do their thinking for them?

Professor Gould does believe that the honey bee's dancing

qualifies as a language. It's nothing, he said, compared with human language, but in comparison with what other animals can do, it's impressive.

Linguist Roger Brown once argued that bees don't have a real language because they don't learn theirs, as humans do. But the forager certainly learns the route to the food, even if she encodes it according to a genetically prescribed formula. Gould has also found that with age, bees improve both in their ability to encode and to interpret dances. It may be just a matter of maturation—growth with no learning involved—but it could also be learning, particularly since we know that they have to learn to allow for the fact that the sun moves westward across the sky at the rate of about 15 degrees an hour; for, when bees are raised in a dark basement and then brought above ground, it's a while before they begin to correct for sun drift. They also have a prodigious memory and can apparently remember the location of a food source for a lifetime—the record, for bees who survived over a winter, is 173 days.

Is the bee self-aware? Does it know what it's doing? Gould, who believes that chimps are as self-aware as many people, said, "I think if I knew bees were self-aware, I'd have to stop doing the sort of work I'm doing, because a number of them succumb to my experiments."

All the same, Professor Gould seems to go to a lot of trouble to see that his bees survive as often as possible. He also obviously has reservations about just how far self-awareness might go in a bee. What impresses him always, he said, is the chaos in the hive. For example, when a bee is trying to carry a dead colleague out of the hive, it will often walk for minutes in all kinds of wrong directions until finally it somehow winds up in front of the hive entrance; yet when it watches a dance, it goes straight to the entrance. What it's doing walking around like that, said Gould, isn't clear. "Bees are not reliable except statistically. If you average ten thousand bees, you get beautiful behavior. If you look at just one bee, you get chaos."

11.

Of Courting Crickets and Chorusing Frogs

The King's daughter began to cry, for she was afraid of the cold frog which she did not like to touch, and which was now to sleep in her pretty, clean little bed. But the King grew angry and said: "He who helped you when you were in trouble ought not afterwards to be despised by you." So she took hold of the frog with two fingers, carried him upstairs, and put him in a corner. But when she was in bed he crept to her and said: "I am tired, I want to sleep as well as you, lift me up or I will tell your father." At this she was terribly angry, and took him up and threw him with all her might against the wall. "Now, will you be quiet, odious frog," said she. But when he fell down he was no frog but a king's son with kind and beautiful eyes.

—"THE FROG-KING,"
by The Brothers Grimm

On a crisp September morning in 1976, I stood in the roadway outside the airport terminal in Ithaca, New York. As the jet I had come on screamed back up into the sky, I surveyed the unprepossessing group of buildings across the road from the airport: long, low, some of them almost barnlike, they were an outpost of Cornell University. It seemed an unlikely place to find Cornell's Langmuir Lab (the part of it concerned with neurobiology), and yet there it was.

I had come to see Robert Capranica and Ronald Hoy, two Cornell scientists who study signaling with sounds. Hoy's animal is the cricket, while Capranica works with frogs, but they share a neurological bias: they're interested not only in the sounds the animal makes and what they mean, but also in the way they're perceived and then processed by the nervous system.

I was particularly interested in crickets because I thought they might provide a kind of paradigm. The messages they exchange are similar in some ways to those of frogs, birds, and other higher animals, and yet they're much simpler. Learning about crickets, then, seemed like a good way to bone up on the basics of auditory communication.

Professor Hoy turned out to be a hospitable and energetic young man who spoke of his research with contagious enthusiasm. On top of a filing cabinet in his office, he had a collection of small wooden cricket cages from the Far East, where people sometimes keep crickets, as we do canaries, for their singing. However, the insects Professor Hoy works with aren't housed in those pretty little cages but are kept, instead, in terrariums, in a small room that's rather like any stockroom, except that inside it sounds like a summer night deep in the country.

Professor Hoy became interested in crickets because, as I had suspected, theirs is one of the simplest sound-communication systems. In fact, crickets themselves are relatively simple creatures, with fewer nerves to carry sounds from ears to brain and fewer cells in the brain itself, and this makes it easier to study them neurologically.

Male crickets chirp (the females are entirely mute) by rubbing one fore wing across the other, and they actually sing several different songs. Most of the pleasant din they create is produced by males broadcasting a calling song from the safety of their own territories. To other males, the song announces that this piece of ground is spoken for and will be defended if necessary. To females, it's a mating cry, and it draws them like a magnet. Males also sing an aggression song while they are engaged in combat, and this is somewhat different from the calling song. In addition, they tick quietly after they have lured a female into their territories, a sound which seems to play some part in courtship.

There are many different species of crickets, and though the aggression and ticking songs are quite similar from one kind to the next, the calling song is distinctive: it's the signature of the species—the way females tell, in the darkness of a warm summer evening that throbs with the calling of a multitude of crickets, which ones are males of their own kind. Female crickets may be mute, but, as Professor Hoy put it, they vote with their feet: when they hear the calling song of their species, they head out in that direction, whether the song is being produced by a bona fide male cricket or by a tape recorder operated by a research assistant. In fact, a female cricket drawn to a chirping loudspeaker will often climb right up onto it, or she may circle it determinedly, but in either case she's apt to make actual physical contact with it—which makes it beautifully simple for watching humans to tell which sounds attract her and to score her response.

How *does* a female single out the call of her kind? It turns out that the temporal pattern of the song is crucial, though the pitch at which it's sung may also be significant. The importance of the pattern was demonstrated some years ago by an experimenter who used electronic equipment to create artificial cricket songs. He found that he could tinker with the sound itself—replacing a four-chirp trill, for example, with a steady tone that lasted about the same length of time—and the song was still almost as attractive to females; but when he fiddled with the timing of the song—the length of the trill, perhaps, or the length of the interval between trills—most of the female crickets no longer headed amorously for his loudspeaker.

The calling song is born into crickets; they don't have to learn it. Since they live about two months, with no overlap between generations, that stands to reason; but there's also laboratory proof that the song is innate. Fertilized eggs have been taken from females and raised, egg by egg, in individual isolation, so that there was no chance the insects could ever have heard the sounds made by their species. The young males went through the usual nine to eleven molts, each time shedding their outer envelopes to have more room to grow, and then, as crickets do, after the very last molt, they began to sing—and their songs were absolutely normal.

Since the calling song *is* prescribed in the genes, Hoy set out a few years ago to find out what kind of song hybrid crickets would sing. He cross-bred two species of Australian field crickets, *Teleo-*

gryllus oceanicus and *Teleogryllus commodus*, and found that their offspring produced calls that were intermediate in their characteristics between those of the parents: for example, the hybrids waited longer between chirps than one parent did and less time than the other.

To find out how hybrid females responded, Professor Hoy used an ingenious but simple apparatus, a lightweight styrofoam Y-maze that looked rather like a donut with another half-donut stuck on top, spanning it like an arch. To a cricket circumnavigating it, this contraption would offer three curved paths connected by two choice points, at which she could turn either right or left.

Because it would be difficult to persuade a cricket to stay on the Y-maze for very long, Hoy tethered the insect, using a toothpick and wax to suspend her from an overhead support so that she hung in midair. Then he presented her with the maze, and since she evidently didn't like dangling, she gripped it with her feet. Usually she simply hung there, maze and all, more or less stationary, until he began to play cricket songs to her from a loudspeaker placed either to her left or to her right. Then she began to move her feet as if walking along the ground, which rotated the maze backward beneath her. Soon she came to one of the choice points, and Hoy could record whether she turned toward the sound of the cricket calls or away from it.

He found that hybrid virgin females not only preferred the song of a hybrid male to that of either parent, but they even preferred the song of a cricket who was a product of the same kind of cross-breeding. If a female had a mother who was a *T. commodus* and a father who was a *T. oceanicus*, that was what she wanted in a male, in preference to one who had a *T. oceanicus* for a mother and a *T. commodus* father. This surprised Professor Hoy a bit, since the songs of the two kinds of hybrids differ only slightly.

But what it suggested to him was that the same set of genes may somehow provide both male and female with a pattern generator somewhere in the auditory nervous system, a generator which produces the sequence of wing movements by which the male calls, but which exists in the female only as a kind of plan against which she can match the songs she hears. How this hypothesized pattern generator could do two such different things is hard to imagine, but scientists have long suspected that some such mechanism (it's called a *template*) exists in birds. In Professor Hoy's analogy, it's as if both male and female come equipped with an internal tape recording of the song of their species—a racial memory, if you will; but in the male, the recording is hooked up to a loudspeaker, while in the female, it plays silently, and she uses it only to judge what she hears. This kind of match-up of internal systems is not unprecedented: not only do male crickets chirp faster in warm weather, but when the temperature rises, females show a preference for songs with a faster tempo. (Some humans can actually estimate the temperature by listening to crickets.)

In the other major thrust of his research, Professor Hoy is trying to find out where and how the female cricket encodes the pattern of the call of her kind. He is, in effect, looking for a mating-call detector within the insect. He conducts his research by playing cricket songs to a female and then recording with a fine electrode the activity of single cells in her auditory nervous system and in her brain. Eventually he'd like to be able to make a kind of wiring diagram to show how a cricket processes sound—a diagram that could be a much-simplified model of the way frogs, birds, and even humans do the same things.

On his desk at the Langmuir Lab, Bob Capranica keeps a carved wooden frog and a preserved specimen, a kind of frog-shell, light as *papier mâché*, given him by a friend. The preserved frog, braced belligerently on bowed front legs, stares back at the world with ancient, goggling, amphibian eyes. It's familiar, homely, and somehow amusing, and looking at it, I couldn't help wondering how a scientist would come to choose frogs to study.

Professor Capranica is tall, rangy, and pleasantly low-key, and it turned out that he was drawn to anurans—frogs and toads—because they're not only very vocal, but they're also, like crickets, relatively simple animals. (Another researcher once said to me, half apologetically, that "frogs are not very fancy.") For example, the anuran's eardrum is actually set into the surface of its head, thus doing away entirely with the outer ear, and its middle ear, nervous system, and brain are much less elaborate than those of mammals. The anuran, or salentian, also produces a small set of simple signals that don't grade into one another, so there are no in-betweens to interpret. All this is important, since Capranica's goal is an ambitious one: he hopes that within his lifetime, we will understand how the frog perceives sound. "That's what most of us in neurophysiology are interested in," he said. "We hope to learn how the nervous system perceives the real world."

It's a fascinating area to work in, since there is probably no nervous system in existence that takes in the world exactly as it is. All creatures have some sensory limitations. In fact, to study any species of animal, a scientist must start by asking: what can it perceive? What is the range of its hearing? What colors can it see? Many animals take in their surroundings through senses so different from the ones humans rely on that the world must seem an entirely different place to them than it does to us. For some, it's totally dark; for others, entirely silent; while to many, it's noisy with smells, a jumble of meaningful odors. The frog is a particularly interesting creature to study because its eyes and ears do a heavy editing job on reality—compared to human senses, they tell its brain very little about what's going on out there, though they do reveal a great deal about the particular events that are of vital interest to a frog.

One of Professor Capranica's more recent studies actually suggests that in at least one species of frog, male and female hear the world in quite different ways. The animal involved is the coqui, a Puerto Rican tree frog so common that it's almost the representative animal of the island—ceramic replicas are common in gift shops all over the territory. It's a very pretty animal, brown with white markings and a large head, and it was named for its call, which sounds to human ears like "ko-kee." The first note is a pure tone, steady at about 1200 hertz (a hertz is a measure of the frequency of a sound, which we experience as its pitch). The second note, also a pure tone, coasts upward from about 1800 to 2200 hertz.

To Capranica, the coqui is particularly interesting because it's very different from North American frogs. For one thing, the males are highly territorial. They take up residence in a tree or dig a hole, and then they sit and call steadily from sunset until about midnight. As with crickets, the call serves as a warning to other males and is simultaneously a mating cry. A female attracted by the call will climb the tree or enter the nest-hole to mate, laying just six to ten eggs—as opposed to the thousands of eggs laid by many North American frogs. Afterward, she departs to find another male and to lay more eggs.

One reason the female can afford to be so parsimonious with

her eggs is that the male actually incubates them, sitting on them to keep them moist. The tadpole stage takes place within the egg, and so after three to seven weeks, they hatch as fully formed, miniature froglets, which remain with their father for a few weeks and then disperse so that he is free to breed again. For the coqui, the mating season lasts eleven months of the year (they don't hibernate), while for some North American species, it's over in just a few days.

Though most North American frogs and toads produce a call that's just a single note repeated over and over again, many tropical species sing two notes as the coqui does, and Professor Capranica wanted to find out why. In 1973 and again in 1974, he went to Puerto Rico with a colleague, Peter Narins. They were equipped with loudspeakers and a portable sound synthesizer, a kind of dial-a-mating call gadget: with this machine, knowing the pitch and pattern of a call, they could simply dial it in and reproduce it artificially.

"Normally, in the rain forest, males are spaced three to four meters apart, and they call every three or four seconds," Professor Capranica recalled. "So we took our loudspeaker, put it near a male, and played him a recording of the natural call. Soon he began to synchronize with it and to call back immediately after the 'co' note; he also dropped the 'qui' from his own call. When we played the synthesized 'qui' note for him, he ignored it and went on calling at his own rate, but as soon as we changed to the synthetic 'co' note he answered again. Observing in the field, we saw that when two males got close together, they stopped calling 'coqui' and began to 'co' at each other, and then they would generally back off. So the 'co' note serves to space the males out, and they pay no attention to the 'qui.'

"Next we studied the females, using the same three sounds—the natural call and the two synthesized elements. To do this kind of study, you take two loudspeakers and put them about four meters apart, and you play a different sound through each. You release a female frog midway between them, turn on one or both, and see which way she goes. If she's responsive, she will hop over to one loudspeaker and actually crawl around on it. Anyway, once she's made her choice, you take her back to the middle and repeat the procedure, and you do this over and over again."

It turned out that the female was the opposite of the male: she was attracted by the natural call and by the "qui" note, and she ignored the "co." Back at the lab in Ithaca, using frogs they brought home, Capranica and Narins recorded from the auditory nerve and compared the sensitivity of male and female. They found that the male's ear is pretuned to resonate with the "co" note, while the female is pretuned to the "qui." Where the cricket broadcasts a single song that has a different significance for male and female listeners, the coqui's two-note call is actually perceived differently, depending on whether the listening frog is male or female. As Bob Capranica put it, "It's kind of a neat signaling system."

"It also serves to emphasize what has happened in frogs," he told me. "They're only interested in a small number of sounds, so rather than transmit all incoming sound information and let the brain interpret it, the frog's ear begins to filter it immediately. Humans, on the other hand, hear most sounds in the environment, though we're poor on low frequencies and don't pick up anything ultrasonic.

"For the frog, this filtering is an economy—it means the brain has a simpler task. It hears sounds that include the mating call, the noises of predators and of some other species, but that's all it hears. The frog can get by, then, with a simpler brain, with a smaller number of neurons devoted to sound communication."

I had asked Professor Capranica whether I could see his lab, so at that point, we broke off and he gave me a guided tour, beginning with the humid, closetlike room where his frogs live in terrariums of all sizes, set out on shelves and on the floor. There was a constant, cheerful chirping from a barrel in the corner marked "Crickets," but when I asked whether these were more of Hoy's experimental animals, Capranica explained that, on the contrary, they were simply frog food. As we made a circuit of the room, lifting off terrarium covers, I saw tiny green toads as bright as jewels; big, homely, mud-colored toads; and frogs of all sizes. Several times there was a mighty splash as a bullfrog dove into water.

In another room was the hardware Capranica uses to study frog perceptions, a whole wall of push buttons, dials, lights, and reels of tape. The actual work with animals is done on a table in an adjoining soundproof room. The frog or bird or whatever is anesthetized and then equipped with a pair of earphones. Hair-thin electrodes are used to contact different regions of the auditory nerves, or perhaps the brain. The machinery in the outer room then goes into action, playing selected sounds to the unconscious animal through the earphones, while researchers watch for a burst of bio-electricity that would indicate that the sound impulse is being channeled through the area where the electrodes are.

Back in Capranica's office again, we talked about North American frogs he has studied; the bullfrog was actually his particular animal for years. The sexual activities of Stateside frogs are rather different from those of the coqui. Breeding generally takes place on damp nights in spring or summer, when males congregate in ponds or swamps and set up a clamor. Sometimes a dozen or more different kinds of frogs chorus from the same pond, and the sound waves spilling out are enormously complex. Somehow, in all that din, incoming females home in on the call of their species.

Having selected a mate, the female frog swims or hops right up to him. He, however, may be so busy calling that she must actually bump him or jump on his back to get his attention. He is, in fact, a fervid but indiscriminate lover, since he will grasp anything that comes near him that approximates in size and shape a female of his kind. Occasionally he gets a grip on another male by mistake, at which point the frog who is being molested vibrates his body and lets out a croak called a release call; apprised of his mistake, the first male then lets go.

Some frogs produce only one call—a few, in fact, are mute—but others have a small repertoire. Professor Capranica discovered that the bullfrog vocabulary runs to seven distinct vocalizations. Besides the release croak, there is a short, loud grunt of warning and a high-pitched scream of distress. Because the bullfrog, like the coqui, is territorial, there are also at least three different territorial calls used to warn off intruders, and then there is the familiar, deep "jugorum" of the male, heard mainly during the breeding season.

It would seem logical to conclude that "jugorum" is the bullfrog's equivalent of the cricket's calling song, a vocalization designed simultaneously to warn off other males and to attract females; and this *is* true of many North American frogs. However, though it does provoke a reply from other males in the laboratory, "jugorum" leaves females singularly unmoved. There is, in fact, no proof that the presumed mating call (if that's what it is) actually attracts females in any of the species that belong to the frog family called *Rana*—a family that includes the bullfrog, green frog, and leopard frog. However, Professor Capranica explained that experimental tests have usually been carried out on dry land, and since these are animals that normally mate in water, that could be the problem.

The exact function of the frog breeding chorus is also still a moot point, for the unfancy frog has presented some knotty problems to researchers. Again, logic would suggest that the chorus serves not only the help females distinguish males of their own kind, but also to help both sexes locate the communal breeding pond; otherwise why would evolution have provided so many frogs and toads with such loud, clear voices? However, there are a few species of anurans that are completely silent, though they still congregate to breed, and others migrate to the pond in silence and then wait out a long prespawning period before they begin to chorus and to mate. There is some evidence that frogs use their sense of smell to find their home ponds; perhaps when they migrate to a breeding pond, they also follow their noses.

What is beyond question is that the mating call is extremely important to the frog, for its ear is pretuned to that frequency. The bullfrog's "jugorum," for example, is a complicated sound, not a pure tone at all but a blend of different frequencies—like a fist or forearm thumping down on a whole clump of piano keys rather than like a single note. However, some of those keys are played louder than others—a few of the low notes and a few of the high ones are projected with more energy than are the rest of the notes that make up the sound. Inside the bullfrog's ear are two papillae or nipple-shaped projections, one particularly sensitive to the low-pitched elements in the mating call, the other to the higher ones; and somewhere farther on in the frog's auditory system, Capranica believes, there must be a mating-call detector, a neuron or group of neurons which reacts when it gets a message from both papillae at once but not when it only hears from one. This, then, may be how the animal distinguishes, from the welter of sound around it, the call of its kind. Presumably, somewhere there is also a warning-call detector, a release-croak detector, and so on.

"If I could figure out how the bullfrog perceives a mating call," Capranica told me, "it might be a model for how humans recognize sounds. Somewhere in your brain you discriminate between the sound of the word 'hello' and that of 'good-bye.' The question is how."

It seemed appropriate to ask at this point whether the unfancy frog has its vocalizations prescribed by its genes, as the cricket does. It appears the answer is yes and no. On the one hand, hybrid frogs, like hybrid crickets, produce vocalizations intermediate between those of their parents. On the other hand, some features of frog calls seem to be learned. There are slight differences from one animal to the next, which probably means that one frog can recognize another frog by its voice. Capranica recalled, "When I was studying bullfrogs, I could go out into the field at night and identify fourteen or fifteen males, old friends, by their voices; and if I could tell the difference, the frogs probably could, too."

Frogs also develop local dialects, which couldn't happen if every aspect of their call were inherited. A New Jersey cricket frog, clicking away very much as crickets do, makes a sound slightly different than a South Dakota cricket frog does. Furthermore, a female cricket frog born and raised in New Jersey, if she is serenaded simultaneously with tape recordings of both dialects, will head for the loudspeaker with the New Jersey accent.

Another phenomenon that Capranica is currently interested in is synchronous calling. Different species of frogs, he said, use different strategies in the breeding chorus. In some species, the males call independently and seem to ignore the sound of neighbors. In other species, the animals synchronize—one calls and a neighbor calls immediately afterward; then there's a pause and the same sequence happens again. Much more rarely, frog number one begins calling, then, after a while, frog number two joins in, but it anticipates and actually sounds off a fraction of a second before frog number one does. Because there must be some reason for each species to adopt the chorus strategy it uses, Capranica would like to know what advantage each offers.

"Anticipation is the real puzzle," he said.

I asked whether it could have anything to do with competition for dominance, but he explained that most species don't have a dominance hierarchy; some live only a few years, and that's probably not long enough for a dominance hierarchy to be a useful way of structuring relationships. Bullfrogs, however, are different. Not only are they territorial, but the older, larger males are dominant and often surrounded by smaller, satellite males. But then a bullfrog may live about fifteen years.

"Amphibians are often not as short-lived as people think," Professor Capranica explained. Some toads have been reported to live thirty years or more, and salamanders may live even longer. The smaller the species, in general, the shorter its life span because—like fish—most amphibians continue to grow throughout their lives, though they grow more slowly as they become older. The larger the bullfrog, then, the older it is.

"One thing I'm disturbed about is the way we're decimating the frog population," Capranica said. "There used to be lots of bullfrogs in this part of the country, but we've wiped out large numbers of them between DDT, which causes the female to produce both eggs that don't survive and abnormal tadpoles, and the demand for frogs' legs. People don't realize when they sit down to eat frogs' legs that if it's a good-sized bullfrog, they may be eating an animal that's fifteen years old. And this species can't come back easily: it takes four years or so before the Northern bullfrog is old enough to reproduce, so if you practically wipe them out, it takes a long time before they can come back.

"The bullfrog to me is the equivalent of the bald eagle," he concluded. "It's *the* American frog."

12.

Translating Bird Song

The two swans flew high and fast, ten thousand feet above the earth. They arrived at last at the little pond in the wilderness where Louis had been hatched. This was his dream—to return with his love to the place in Canada where he had first seen the light of day. . . . Serena was enchanted. They were in love. It was spring.

—THE TRUMPET OF THE SWAN,
by E. B. White

The idea that animals are rather like soft machines, programmed by their genes to behave in predictable and inflexible ways, came primarily from studies of birds.

For instance, I've read in textbooks that the sight of anything red will "release" threatening behavior in a male robin—that it will energetically attack a bundle of red feathers dropped within its territory, while ignoring a stuffed robin that doesn't have a red breast. Furthermore, the parent robin devotedly feeding its young is said to be simply responding in a stereotyped way to a stimulus that includes both the reddish insides of the beaks of the young birds and the nest itself, since it will ignore its young completely—no matter how noisily hungry they are—if they're outside the nest, even if they're only inches away. The signals exchanged by birds are sometimes said to work in a similar way: like a key engaging a lock, the sound of a rival singing within its territory supposedly turns on an aggressive response in a male bird.

However, we know now that although some signals of some species are indeed fixed and predictable, others are not, and, in fact, today the scientists who study bird communication are finding it so complex that, inevitably, some are beginning to make comparisons with human language. Especially in the matter of bird song, the parallels to language *are* quite startling.

The Rockefeller University Field Research Center in Millbrook, New York, is deep in the countryside. Once the gate house to a large estate, it's a big, handsome place that looks more like a ski chalet than like what it is: one of the major American laboratories for studying bird song. On the inside, the building is a regular warren of short corridors and large, sunny offices. There are walls as thick as those of a Cotswold cottage, and some of the doors are shaped like archways. The ambience is part graduate school, part country estate.

What I remember most vividly, though, is the bird lab on the ground floor. I was shown around the lab by Roberta Pickard, a research assistant, and when she opened the door of one particular room, a sound like steam escaping rushed out at us. The sound is called white noise, and it's used to make sure that no canary arias, no vagrant scrap of swamp-sparrow song, can reach the ears of the birds inside the room, some of whom have been raised from the egg in white noise without ever hearing so much as a peep or a chirp from any kind of bird.

Inside the room, dozens of boxes sat side by side on utility shelves. They looked rather like wall ovens, each with a little window in the front, but they were actually soundproof containers. When Ms. Pickard opened one of them, I saw that it held a large microphone and two long-billed marsh wrens—small, slim, racy-looking brown birds. They were in adjoining cages, and the wren on the right was a young male who hadn't yet begun to sing, while the adult wren on the left was his song tutor. The latter was actually doing double duty, for the microphone piped his lessons on to yet another young bird alone in a cage a few boxes away. Eventually it would be possible to compare the song of the bird who was able to see the tutor with that of the bird who was only able to hear him, to find out if there were differences. But what particularly intrigued me was the fact that the tutor himself originally learned his song not from an older bird but from a tape recording.

More than twenty years ago, scientists in England first proved—by raising young chaffinches in soundproof isolation—that all song birds aren't born knowing exactly what to sing. Peter Marler, the Rockefeller University professor I had come to Millbrook to interview, was involved in some of those early experiments at Cambridge University, and after he moved to the United States in 1957, he added a great deal to what we know about song learning through extensive studies of California's white crowned sparrow. A solid, thoughtful, slow-spoken Englishman, he's immensely respected in his field.

Professor Marler was in conference when I arrived at Millbrook, so while I waited to talk to him, I rapidly reviewed some of what I'd already learned about bird song. There were, I had decided, seven principal facts one had to know to understand the basics of bird song.

1. In most species, only the male bird sings. As with the cricket, his song identifies him as a member of his species and attracts females. Simultaneously, it warns other males away from his territory.

2. It's also his personal signature. Scientists have found that in many species, one bird can recognize another by his song.

3. Many birds also sing in dialect: all those living within a given area produce the same variations on the basic theme.

4. For some species, there's a critical period for learning to sing. If, during the first few weeks or months of life, scientists and soundproof boxes prevent them from hearing the song of their kind, they grow up to sing abnormally no matter how often they hear that song *after* the critical period is over.

5. A young bird raised in a soundproof box will learn normal song quite readily if he's allowed to listen to recordings of wild birds singing. However, most will learn only the song of their own species, ignoring those of other birds, even if that's all they ever hear.

6. The young male bird, hatched in spring or summer, doesn't begin to sing until his hormones prompt him in the spring of the

following year. In the meantime, he practices with something called *subsong,* which is soft and rambling and only gradually crystallizes into the themes of his species.

7. An adult bird that is deafened surgically continues to sing quite normally. If the same operation is performed on a young bird who hasn't yet begun to sing—even if he's old enough to have heard and learned his song—when he grows up, he won't sing at all but instead will make strange, insectlike sounds.

Years ago Professor Marler took all these separate facts and made sense of them with a single theory. He suggested that the male song bird is equipped from birth with a *template,* a rough plan of the song of his species. Unlike the cricket, which is born knowing every note and nuance of its calling song, the young bird often knows only enough to insure that as he develops, he'll tune in to and learn the themes of his own kind, rather than those of some other species. As he matures, then, he learns from his elders and improves on the template by adding details, including a local dialect and some personal touches. If he's raised in soundproof isolation, of course, he can't learn these details, which is why he sounds somewhat abnormal; and if he's deafened before he has come into song, he can't even approximate the anthem of his kind, because he hasn't been able to listen to his own singing and match it to his template by practicing with subsong.

Not all birds learn to sing in just this way. For some, the song plan is complete at birth, or very nearly, just as it is with crickets. Others use a different learning strategy. The bullfinch, for example, simply acquires the song of the male who tends him, whether that male is a bullfinch or not. But what's important about birds such as the white crowned sparrow and the chaffinch is that there are intriguing parallels between the way they they learn to sing and the way human infants learn to talk.

Song learning in birds is, then, a complicated but important subject, and the setup at Millbrook is quite elaborate. When Professor Marler took me around, he began by showing me a formidable array of sound-analyzing equipment.

The problem in studying animal sounds has always been that humans miss so much. Most of the time, we're not aware of the spread of frequencies in a sound or the subtler shifts in pitch and loudness. We can't tell just by listening when a trill in a bird's song is one syllable shorter than usual. The solution has been to freeze sounds by making them visible.

Dr. Marler explained that within about the last five years, there has been a breakthrough in techniques for analyzing sound. For a long time, scientists were dependent mainly on the spectrograph, a machine that graphs short swatches of sound, plotting the frequency (or pitch) along the vertical axis, with time along the horizontal one. Marler showed me a spectrogram of the trills of a swamp sparrow's song, and it looked a bit like a musical score, with the notes of the music drawn as a row of slanting, upside-down exclamation points. I could easily see the way the pitch shifted upward as the lines did.

The spectrograph, however, has its disadvantages. It's a small machine topped with a revolving drum that's wrapped with a strip of paper. You play bird song into it, and as soon as you hear sounds you want to analyze, you pull a switch. Several minutes later, the machine traces out on the revolving drum a spectrogram of the 2.4 seconds of sound that preceded the pulling of the switch. Patching together a picture of a stretch of sound longer than 2.4 seconds is obviously a tedious business. About five years ago, life became much easier for bioacousticians (those who study animal sounds) when "real time" sound-spectrum analyzers became available. This new machine can provide a continuous display almost instantaneously on a screen or within minutes as a permanent photographic copy.

The research team at Millbrook is also trying to design a machine that can synthesize bird songs. Ron Hoy already has an artificial cricket, and Bob Capranica can dial up a frog's mating call, but crickets and frogs make simple sounds in comparison to bird songs, and it will take some fairly sophisticated equipment, including a computer, to mimic birds. Once the Millbrook synthesizer is perfected, Marler will be able to copy songs and then to alter them at will in subtle ways to find out which features are significant to birds.

I had arrived at Millbrook at a relatively quiet time. It was March, and so it was too early for last year's hand-raised birds to be coming into song. Hence there were no answers yet to last year's questions—questions which were asked by rearing the birds in particular ways: by allowing one bird to see his song tutor, for example, while another could only hear him. It was also too early for fieldwork or for collecting wild birds to raise in the lab.

Hand-raising birds is a difficult task, since newly hatched song birds are incredibly fragile creatures—some are no larger than a thimble. When I asked Professor Marler how it's done, he explained that "it's kind of a trying phase of the research." Sometimes the isolation experiments are done with wild birds taken from their nests when they're just a few days old. At other times, eggs themselves are taken and given to canaries to incubate and rear. White noise is piped in to make sure a nestling never hears the song of its foster father or of any other bird.

Even with canary foster parents to help out, however, raising young birds is a demanding job, for though the canaries do feed their young charges, they don't feed them enough. Humans have to supplement their diets, using surgical tweezers to poke the food down the tiny, gaping mouth, and at least in the beginning, this has to be done every fifteen or twenty minutes from dawn until dusk. Then there is the chore of concocting the bird food for both nestlings and adults. The day I was at Millbrook, Dr. Donald Kroodsma was getting ready to mix up another batch of it for the long-billed marsh wrens he's studying: he combines chopped liver, cottage cheese, ground beef, ground-up insect larvae, vitamins, and other ingredients until he has eighty pounds of gray goo—enough to feed thirty to thirty-five wrens for about three weeks.

Dr. Kroodsma, a tall, fair-haired young man, has been studying the responses of female birds, and in a small, notably quiet room (the females don't sing, of course), I saw some of the swamp sparrows he has been working with. Skinny little brown birds with pale, brown-streaked breasts, each was housed in a cage that was equipped with a wire nest-cup, a gadget that looked like a strainer without a handle. A packet of short, white strings had also been supplied. It seems that researchers in England have demonstrated with canaries that when a female is provided with a nest cup and a bundle of strings, the number of strings she pulls out for nest building in any given day correlates nicely with her current reproductive development. Dr. Kroodsma himself recently showed, again with canaries, that the rate at which a female develops sexually can be influenced by the sound of a male's song. Kroodsma serenaded females with normal song and with an artificially simplified version. Using the number of strings pulled and eggs laid as measures of sexual development, he found that the females who heard the simplified song matured more slowly.

Kroodsma is not trying to repeat the experiment with swamp sparrows to see whether other birds react as canaries do. He told me he was delighted to be working again with a "real bird" as opposed to a canary, which has been domesticated for so long that "we're not sure what has happened to it." However, the string-pulling format has never been used before with wild birds, and no one knows whether it will work. So far, Kroodsma's birds had dropped bits of paper and chunks of food into the nest cups and had occasionally hopped in themselves, but they had steadfastly ignored the bundles of string.

Female marsh wrens housed in the same room would eventually take part in a different study. In a procedure very similar to the way Bob Capranica tests female frogs, Kroodsma planned to place a young wren in a very long, narrow cage with a loudspeaker positioned at each end. He would then play for her the sweet, slightly raspy song of her kind, first from one speaker and then from the other, to see which she would head for—assuming that she was cooperative enough to head for either. In the wild, male long-billed marsh wrens often sing back and forth to one another. The first bird will render song A from his repertoire and the second replies with song A. The first then produces song B and is answered with B, and so on, up to as many as one hundred themes. Dr. Kroodsma hopes to find out how female marsh wrens choose a mate—whether they generally pick the male who leads off, for example, or the one who sings loudest, or whether there is another determining factor.

In the meantime, Kroodsma's experimental birds were being treated to an early spring. Though it was only March, they were being given light for fifteen hours a day to hurry along their reproductive development, since the experiments must be completed before the start of the fieldwork season. Don Kroodsma's new field project involved studying the mating habits of swamp sparrows. From the middle of April until the middle of the summer, he and his research assistant would be spending most of their days in hip boots, wading through a marsh, equipped with binoculars for observing and recording the activities of sixty to one hundred mated pairs. The birds have been banded so that, from a distance, each can be identified by the colors of the bands it wears on its legs. Five to ten percent of swamp-sparrow males are bigamists, and Dr. Kroodsma is interested in what makes one bird able to attract two mates when most have only one. Is it his singing, the size of his territory, or something else altogether?

That day at Millbrook, Professor Marler had just finished writing up an experiment of his own from the previous year that provided some surprising new insights into the template mechanism. Apparently it may be something extremely simple that insures that the young bird learns the song of his kind.

In this experiment, Marler worked with swamp sparrows and their larger, plumper relatives, song sparrows. These two species live within earshot of one another in the area around Millbrook, and they sing very different songs, the swamp sparrow producing simple trills while the song sparrow's song is complex and varied, a delightful jumble of whistles and trills.

To find out more precisely what it is that prevents a swamp sparrow from growing up to sound like a song sparrow (and vice versa), Professor Marler raised young males of both species in soundproof isolation and tutored them with artificial songs. Just as another experimenter once tinkered with cricket songs to find out which elements meant something to the cricket (see Chapter 11), Marler, in this more complex study, tried out variations on the birds' basic themes. He fully expected that—as with the earlier cricket experiment—it would turn out that the temporal pattern of the songs was important, since this was the most obvious difference between them. Swamp sparrows, then, should learn simple trills that he played for them, and song sparrows should pick up the complex patterns similar to their own.

However, it was also possible that there was something distinctive about the *syllables* themselves. (Syllables are identical units of sound that are repeated in a string—for example, each pulse of a trill is a syllable.) So he decided—almost as an afterthought, he said—to create duplicates of each pattern. Thus he constructed a swamp-sparrow-type trill using a swamp-sparrow syllable, and another with a song-sparrow syllable; he also reproduced song-sparrow themes with syllables from both species. In the absence of a synthesizer, he managed this by extracting syllables from tapes of wild sparrows singing and then rerecording them according to the patterns he'd designed.

Each young male was presented with one set of songs made from swamp-sparrow syllables and a second set made from song-sparrow syllables. To Marler's surprise, the birds completely rejected the songs constructed from alien syllables and learned only the ones built from syllables sung by their own species. Obviously it was some feature of the syllable itself that insured that they learned only the song of their kind. But then the birds went on to do something even more surprising: a swamp sparrow, for example, would extract a swamp-sparrow syllable from a pattern completely unlike any that swamp sparrows ever sing, and he would trill it in the temporal pattern typical of his species. So he *did* somewhere have the information necessary to produce that pattern.

This has started Professor Marler wondering whether some birds may not refer to the template at more than one stage in their development. Perhaps that distinctive syllable exists as something very elementary that simply directs the bird's attention to the right song during his critical learning period. Later, when he's ready to start uttering songs himself, he may refer to his inborn knowledge about the temporal pattern of the song.

The template concept is similar in some ways to the ethological concept of innate releasing mechanisms. In one case, the bird is said to be born endowed with a rough song plan; in the other, it's thought to be born with a readiness to respond in a specific way, for example, to the sight of reddish beaks inside something shaped like a nest. However, Professor Marler believes that the old notion of releasers which work like a key in a lock may be rather removed from reality.

"The importance of releasers," he told me, "may lie not so much in programming the animal as though it were an automaton, but rather in leading it to be selective about what it will pay attention to. Beyond that, a good deal of learning may take place." Marler believes that much of the behavior that ethologists describe as innate may come down to simple, attention-catching mechanisms like the swamp-sparrow syllable, and that such mechanisms may exist to guide the development of human language, too.

"That's what excites me in bird work," he said, "that eventually it may have more general applications."

When one considers the way birds learn to sing and the way babies learn to talk, some of the similarities are so obvious that they're easy to overlook. Just as birds learn the song of their kind and not that of some other species, so babies learn to speak and not to bark or to chirp. And just as some birds learn dialects, humans learn different languages—and, of course, local dialects of those languages. Furthermore, people—like birds—can recognize a neighbor by his or her voice or tell the sex of a stranger just by listening.

In infants, babbling seems to serve the same function that subsong serves for birds: it's a chance for the baby to exercise her vocal equipment and to try to match her own sounds to words she has heard spoken by adults. Early babbling (like early subsong) sounds much the same for all babies, but late babbling (like late subsong) shows evidence of learning: an older Chinese baby sounds different from an older French baby.

The drive to learn to speak or to sing seems to be present in both babies and young birds almost from birth. And just as some birds have a critical period for learning song, many linguists believe there is a critical period for humans—that if a child isn't exposed to language during the early years, he will never learn to speak.

There are parallels, too, in the problems faced by children who become deaf before they have begun to speak and those of young deafened birds, since both are unable to match the sounds they make to sounds they've heard (or to an inherited template). But the most fascinating correspondence of all is the anatomical one: both the ability to speak and the ability to sing reside in the left hemisphere of the brain. And the similarity doesn't stop there. An adult human with left-hemisphere damage often loses the ability to speak or to understand language, though very young humans with similar damage generally grow up to talk quite normally, since the right hemisphere takes over the tasks of the left. In much the same way, if an experimenter cuts the nerves that connect an adult bird's vocal equipment with its left hemisphere, afterward it can sing only the most fragmentary song, while if the same operation is performed on a young bird, it eventually comes into song quite normally.

Marler suggests that humans may also have something very like the template he has postulated for song birds: there may be innate mechanisms that first focus the infant's attention on speech sounds as opposed to nonspeech sounds and that also provide an orderly frame of reference as the baby listens to those around him.

Human speech is not only highly complex as sound communication goes, it's also surprisingly variable. This is something we're not generally aware of, perhaps because we're influenced by written language and tend mentally to alphabetize what we hear. And yet the same letter, embedded in two different words, is acoustically quite different. The *t* sound in "temporary," run through sound-analyzing equipment, doesn't look the same as the *t* in "plate." In fact, according to one estimate, there are ninety possible ways to pronounce *t* if you're English-speaking.

Because of the complexity and variability of speech, and because, as I've said before, infants are never really taught language, it's easy to imagine two children analyzing and interpreting what they listen to in completely different ways. Yet this doesn't happen, and so it seems likely that there are some innate guidelines, similar to the song bird's template.

Linguists studying the phenomenon called *categorical perception* believe they're on the track of some of these guidelines. Apparently it's hard even to pick out features of speech that can be analyzed by a machine, but one thing that *is* measurable is voice-onset time, the time lapse between an implosive sound made with the lips and a voicing sound that's made in the throat. For example, in the syllable *pa*, the interval between the *p* and the *ah* sounds is usually 40 to 50 milliseconds, while in the syllable *ba*, there's almost no interval at all. Presented with a taped, graded series of speech sounds where the voice-onset time moves in 10-millisecond steps from +100 milliseconds to −100 milliseconds (the sounds are produced by a machine for absolute accuracy), the average English-speaking adult hears all the sounds with a voice-onset time longer than 25 milliseconds as *pa* and all those with a shorter interval as *ba.* (It's as if a whole spectrum of pinks and reds were visible only as either pink or red, with no gradations between.)

Furthermore, tests done on infants as young as one month old have shown that they process *pa* and *ba* much the same way adults do. At about the same point that adults hear the syllable change from *pa* to *ba*, babies show by a quickening of interest that the monotonous "voice" they've been listening to has just said something new. This is true even of babies growing up in a tribe in Africa which has no *ba* sound in its language, so that they have never heard *pa* and *ba* contrasted.

However, there is a controversy over how the *pa/ba* threshold should be interpreted. Some linguists favor a simpler explanation than the assumption that it represents a special perceptual mechanism humans have for analyzing speech, and they cite the curious phenomenon called *backward masking*. In backward-masking experiments, psychologists take two sounds—a musical tone followed by a blast of noise—and prepare a tape in which the interval between them steadily decreases until they overlap: the noise, in effect, backs up over the tone until the tone can't be heard. Then they cut and splice the tape so that the tone-and-noise pairs are presented in random order, and they play the new tape for experimental subjects, asking after each pair, "Can you hear the tone?" It turns out that the subjects stop being able to hear the tone a little before the point at which the noise physically encompasses it. In effect, something they haven't heard yet cancels out something they are at that instant listening to. This tends to happen when the interval between the tone and the noise is around 20 milliseconds—just as people begin to hear the *pa/ba* difference at around 20 milliseconds; so some psychologists say that the voice-onset boundary may be nothing more than a distinction between one set of conditions in which the human ear can hear that there is a delay between two sounds, and another set of conditions in which it can't. Professor Marler told me: "There's a good deal of argument about just how compelling the evidence is for a special speech mechanism. The argument is very intense. No one is quite sure how far they can push either interpretation."

I asked Marler what he thought of Noam Chomsky's theory that humans are genetically programmed to develop certain kinds of syntax. He replied, "To my knowledge, Chomsky has never specified his notion in developmental terms. There are those who interpret him as saying that the child has some elaborate preordained concept of the grammar of human language. However, another possibility is that the common rules adults share result from a rather simple, elementary set of initial instructions which don't in themselves represent the complete body of syntactical rules but which provide sufficient instruction to make it very likely that as the child begins to use language it develops in a somewhat predictable way."

Marler believes that "instinctive" behavior will eventually turn out to be less fixed and inflexible than it has seemed to be, and that at any stage of development, what individuals can learn will turn out to be more constrained than we ever thought it was.

Given the fact that human children and young birds, as they mature, have to cope with such different problems, it seems rather strange that there should be so many similarities in the way they learn to speak or to sing. On the other hand, there are too many parallels to be simply a coincidence. Professor Marler believes that there may be a basic set of rules for vocal learning that would hold true for any species with social relationships that depend on complex, learned traits.

I've spent a great deal of time on bird song because of the fascinating resemblances to human language. However, songs are far from the whole story, since birds also call. In fact, there are about 8700 different species of birds in the world today, and although all of them call, only about 5000 species also sing.

A call is much less complex than a song. Usually it's just a single, short sound, which may be repeated a number of times. A song always has at least a few different notes, delivered in an organized pattern, and the pattern itself is what is repeated. Of course, singing also goes with the breeding season and serves to attract a mate and stake a territorial claim, while birds call all year round—to signal danger, distress, or hunger; to announce that they've found food; to bring the flock together; and in other situations. Some birds actually produce different calls for different kinds of danger. Chickens cackle something that sounds like "gogogocock" for danger on the ground and squawk "rehh" for danger from the air. There are bird calls that are understood across species lines: many small birds squeal "seet" when a hawk flies overhead, and at the sound, all birds in the vicinity will take cover.

For a time, scientists believed that even if some songs were learned, calls were probably all prescribed in the genes. Ring doves and domestic chickens, for example, when raised in isolation, produce perfectly normal calls. However, vocalizations seem to develop rather differently in game birds. For one thing, the young are precocial—they're more developed and not nearly as helpless at hatching as young song birds are. Furthermore, family communication begins while the embryo is still in the egg. Duck eggs, for example, generally hatch about twenty-seven days after they are laid, and from the twenty-fourth day on, faint clicking sounds come from the eggs as the unborn ducklings inside clap their bills in response to calls from the mother duck. This clicking seems also to stimulate the tiny birds themselves, so that they hatch sooner and more or less at the same time. And within a short while after hatching, they will approach and follow anything that sounds like a mother duck—no matter what it looks like.

For other nonsinging birds, however, it's a different story. We know now that calls, like songs, are sometimes partly learned and can be distinctive enough for one bird to recognize another. More important, recent studies of the laughing gull suggest that birds who don't sing *can* have repertoires of signals at least as complex as that of song birds, with fascinating—and different—parallels to language.

In fact, it was at a conference on the origins and evolution of language, sponsored by the New York Academy of Science in 1975, that I listened to Professor Colin Beer deliver a paper modestly entitled "Some Complexities in the Communication Behavior of Gulls." Though there were a number of papers given at the conference on primate communication, Beer's was the only one on a nonprimate species, and it represented a bold attempt to bridge the gap between the way linguists approach the study of communication and the way animal behaviorists do.

Professor Beer is a New Zealander. Now at Rutgers University, he does his fieldwork at the Brigantine National Wildlife Refuge in New Jersey, where he studies the laughing gull, a bird found along the southern coasts of the United States and as far north as Maine in the east. In summer, it has a black head with white rings around the eyes, and in any season, it can be recognized by one of its calls, which sounds like derisive laughter: "Yuck-yuck-yuck-yuck."

The adult laughing gull actually has a repertoire of about a dozen different calls, but the long call—the one that sounds like laughter—is the most complicated both in its pattern and in the way it's used. It consists of a string of call notes delivered in three sections. First, there is an opening sequence of short, quick notes; then the bird sounds a string of longer notes spaced further apart; then a final, few "yipping" notes, unevenly spaced and each generally accompanied by a sharp backward toss of the bird's head.

Long calls are given in a wide range of situations, and there are particular variations in them that go with particular circumstances. For example, a courting bird landing beside a potential mate produces an unusually long string of long notes, while an incubating gull, calling to a bird not its mate, abbreviates the long-note section.

Professor Beer found that he himself could recognize the long calls of individual gulls and, not surprisingly, when he played back tapes he'd made, he proved that the gulls could do it, too. The way the initial short-note section was uttered was particularly distinctive, since some gulls rattled it off with machine-gun delivery, while others gave only one or two relatively long short-notes. Beer also found that a gull directing its long call to its chicks addresses them differently from the way it does another adult bird. Calling to chicks, it begins loudly and becomes softer, rather than the other way around. What's more, two-week-old chicks will only approach at the long call of their own parents and only if that call is the version specifically addressed to chicks.

Summing up, Beer wrote that a laughing gull can use the long call to emphatically identify itself and at the same time to convey any of a number of different messages. It can say, for example, "I am your parent—come and get fed," or "I am your mate—let me sit on the eggs," or "I am your prospective mate—come and stay close," or "I am the occupier of this area—get out." To a surprising degree, the long call is open-ended, as human language is.

Beer has also analyzed other laughing-gull calls. His spectrograms show that the birds' whole repertoire of about a dozen different calls is composed from just a few (perhaps six) different types of call notes. In fact, he says one can draw a distinction for laughing gulls between minimum units of sound and minimum units of sense. Linguists, of course, do something similar with language when they talk of phonemes (the smallest significant particles of sound) and morphemes (the smallest meaningful units of speech—"un" as in "undone" is a morpheme). Some laughing-gull notes, such as the two-syllable "ke-hah" and the so-called head-toss note, are also minimum units of sense—alone, they serve as signals. But most have to be strung together to make a call. Sometimes the same note is repeated over and over, as in the copulation call; sometimes several different notes are combined, as in the long call.

What Professor Beer has done in studying laughing gulls is to look at their signaling from a radically different point of view. In the paper he delivered at the origins-of-language conference, he included a brief history of the scientific approach to animal communication. He recalled that in the beginning, communication was viewed as a stereotyped lock-and-key process. As this explanation proved inadequate, scientists began to speak instead of probabilities and to do statistical predictions of the behavior sequences that could be expected. The notion of context was also introduced: it was recognized that the meaning of a signal came partly from the situation in which it occurred—for example, a laughing-gull call that sounds the alarm when given inside the gullery announces a food-find when uttered outside. However, the theory was still not flexible enough to describe some of the more complex communications of animals.

Why has this complexity eluded ethologists? Speculating on the reasons, Beer mentioned "the bogey of anthropomorphism"—the fear many scientists harbor of mistakenly attributing human traits and reactions to animals—plus the notion that animals are genetically programmed machines. As long as we think of animals as machines, Beer said, "the conception of the animal as an active agent using action in the pursuit of ends will probably not even come to mind. The idea of use here entails the notion of intention, and this is too mentalistic a notion for most behaviorists. . . . Hence the suggestion that an animal might use the same signal to express different messages, by varying what accompanies or occurs in sequence with it, has a heretical taint and so will not come to the minds of the pure in science." He noted that the suggestion also bears a suspicious resemblance to a description of language.

Pointing out that the more animal communication is studied,

the more complex it turns out to be, Beer advised that those who compare animal signaling with language, whether to dwell on the apparent differences or the apparent similarities, should be rather tentative about it, "lest tomorrow's discoveries make today's conclusions look silly."

13.

Eavesdropping on Whales

At the first glow of dawn the Little Calf and a companion of his own age have filled their bellies with warm milk. Now they are looking for fun. A vague tingling, an unquiet, grows in the water about them. For a while they pass it off as porpoise talk, a part of the daily world, a murmur as common as the lapping of waves. But ever more clearly a pulsating beat comes through the sea. In unison the whales flex their muscular bodies and surge ahead to gain velocity. In a rush they raise their stubby heads above the water and gaze quickly but intently toward the horizon, right and left. . . . A half-mile away they dimly see the flashing bodies of white-sided dolphins at play . . . hot in pursuit of a low-hulled gray ship. . . .

—THE YEAR OF THE WHALE,
by Victor B. Scheffer

Vast, dark, encrusted with barnacles, the gray whale flowed along at a sedate five or six miles an hour. The year was 1969, and it was the season of the whales in San Diego: between the middle of December and the middle of February, approximately eleven thousand gray whales would pass by, surging southward along the coast toward breeding grounds in warmer waters. This particular animal was traveling alone, and as it swam it knocked, moaned, and bubbled, as gray whales are apt to do when they're migrating. The sounds, however, were inaudible above the surface of the sea.

Suddenly the thin, eerie screams of a pack of killer whales cut across the creature's underwater monologue. The gray whale reacted immediately and, turning, swam directly into a bed of kelp. Scientists believe that killer whales use sonar when they hunt, and kelp, which is full of gaseous bladders, impedes sonar. Though it seems absurd to think of such a large animal hiding, that's exactly what the gray whale did.

Once inside the kelp, it exhaled underwater so that there was no *blow* (the jet of warm air that shoots up every time a whale breathes out). Then it came coyly to the surface and inhaled very quietly. When it went down again it stayed under for a long while. In this fashion, it waited, remaining silent and virtually invisible except that from time to time it would thrust its big gray head up above the kelp, as if for a look around.

However, there was nothing to see, because the screams came not from prowling killer whales but from a navy catamaran anchored just off the kelp bed. The scientists aboard the boat were discovering that by broadcasting recorded killer-whale screams underwater, they could affect the behavior of gray whales almost every time. Sometimes instead of hiding in the kelp, the animals would retreat to the north or swim out to sea, but few failed to react. On the other hand, whales that were not greeted with killer-whale screams continued smoothly on their way, and so did animals to whom the men broadcast random noises or sounds that only resembled the screams.

One of the intriguing things about this playback experiment is that it amounted to communication from one species to another, using the signaling system of a third species. To date, scientists have virtually no idea what killer-whale screams actually signify, and probably gray whales haven't, either: they may be simply recognizable as the voice of *Orcinus orca,* the killer whale. However, Dr. William C. Cummings, who headed the group of scientists involved in the playback experiments, recently began to try to decipher the communication of the *orca,* working with two of the star performers at Sea World in San Diego.

In the fall of 1976, I flew to California to spend two days with Dr. Cummings. He was then a bioacoustician at the Naval Ocean Systems Center (NOSC) in San Diego, though he's now continuing his research as chief scientist at the San Diego Natural History Museum. Cummings is a tall man with rugged features and large, very blue eyes. An expert in the sound production of whales, he began my visit with a tour of the Marine Life Sciences Laboratory at the center; it sprawls along a shelf of land at the foot of a cliff, literally a stone's throw from the Pacific. The sun was hot that day, and I could hear the steady, interminable barking of sea lions as we stuck our heads into several freezers, each the size of a garage, where fish for feeding the center's animals were stored. At the time of my visit, there were about twenty-five animals in residence, mostly porpoises, fur seals, gray seals, and sea lions. One porpoise, alone in a shallow pool, was busy pinging its sonar off targets suspended in the water by experimenters. There were several tall redwood tanks nearby and, climbing a flight of stairs attached to the side of one of them, we looked down on two more porpoises.

"See if they'll let you touch them," Dr. Cummings suggested. "You'll be amazed at how soft the skin of a porpoise is."

But the animals circled warily just out of reach. Well and recently fed, they had no use for me.

Dr. Cummings's office was in a trailer, one of several that gave the laboratory an extemporaneous air. Once we were inside, he began my education in whale signals by playing a tape recording, a kind of audible catalog of marine-mammal sounds, made for the navy's sonar operators, who have to be able to distinguish porpoise sounds, for example, from the noise of an approaching vessel. The tape (one of the navy's "unclassified" teaching aids) was a fascinating medley of moans, howls, squeaks, and whistles—it seemed to encompass almost every imaginable inarticulate sound. I learned that the Weddell seal makes a whistling noise like bombs falling. Pilot whales signal back and forth as they hunt squid with snores and bird calls, and sperm whales sound like a whole squad of carpenters hammering and sawing. The tape ended incongruously with a march played by a navy band.

Next, Dr. Cummings proposed that we look at films of some of the research that has been done to date by scientists interested in whale communication, and so we moved to a conference room in another building. As we sat waiting for the film projector to be set up, I reviewed with him the background reading I'd done.

Whales—or cetaceans, as zoologists call them—seem the most mysterious of animals, incomprehensibly different from humans. True, they're warm-blooded, air-breathing mammals, like humans, but they can dive to incredible depths, far beyond the point at which we get the bends, and can hold their breath for unthinkable lengths of time. It's said that a sperm whale can stay down for an hour. Some cetaceans can plunge to a depth of seven thousand feet, a fact uncovered through locating the sounds they produce.

In evolutionary terms, however, cetacean ancestry is surprisingly mundane. They are sea-going ungulates, most closely related to cloven-hoofed animals such as cattle and sheep. Furthermore, the whale embryo is startlingly like a pig embryo, or, for that matter, a human one. It has a neck, ridges where ears might be, and bumps for legs. This suggests that whales were probably at one time land animals, for in mammals the embryo often seems to recapitulate, as it develops, some of the evolutionary history of the species. In the whale, all these features disappear before birth, but other resemblances persist. Inside the creature's great flippers are bones very like those of the human hand and arm, with, in most cases, five fingers, though some species have no thumb.

In addition, as most people know, cetaceans seem to be as playful and curious as kittens, and some species, at least, are impressively intelligent and have a highly developed social structure. Whales will often come to the rescue of one of their kind, and whale mothers as they give birth are sometimes attended by another female, who takes it upon herself to raise the newborn to the surface for its first great gasp of air.

Many people think of all whales as being of one general grouping, but actually there are two different kinds, the toothed whales and the baleen, and it's thought that they began to evolve along separate lines as much as twenty million years ago. The species that have teeth are mostly small (at least as cetaceans go), except for the mighty sperm whale. This group includes the pilot whale, the beluga, and all the different kinds of porpoises right up to the killer whale, which is the largest of the porpoises. Baleen whales are so called because instead of teeth they have baleen, or whalebone, which looks like an enormous comb fixed to the upper jaw. These creatures eat by gulping in gallons of ocean, then squirting the water out again through the baleen, which holds back all the small, edible morsels. There are ten species in this group, including the humpback and the California gray whale.

Besides size and dentition, there are other, more profound differences between the two kinds of cetaceans, and these are reflected in the sounds they produce. Baleen whales utter low-pitched moans and pulses, often repeated in simple patterns rather like bird song, and Dr. Cummings suggested that any information they exchange would doubtless be relatively simple. Toothed whales, on the other hand, usually have a whole repertoire of complex, high-frequency sounds that can vary a lot in pitch and repetition rate. Some of these sounds are evidently used for communication, while others are part of the animal's sonar system. Though it's communication we're concerned with here, cetacean sonar is too remarkable to skip over completely. In any case, Cummings believes that some of the sounds thought to be used for echolocation may instead serve to communicate; or perhaps they do both simultaneously.

Because porpoises are a manageable size, their sonar has been studied much more extensively than that of other whales. To find its way around in the dim depths of the sea, the porpoise produces trains of high-pitched clicks. When these sounds strike an object, echoes are reflected back for the animal to analyze—and the analytic abilities of a porpoise are truly extraordinary. A blindfolded bottle-nosed dolphin, for example, can locate a half-inch vitamin capsule at the bottom of a pool. It can distinguish a small square of copper from a small square of brass of exactly the same size. It can tell a circle from a triangle. Even more remarkable, it can project its sonar beam in two distinct lobes, each traveling in a different direction and one more intense than the other; it can even switch and make the weaker lobe the stronger, and then change the shape of each beam. Dr. Cummings told me, "We would need two generating mechanisms that together were the size of a small room to be able to produce and focus a sound beam as well as a porpoise can."

We know a great deal more about cetacean sonar than we do about cetacean communication, which is harder to study. Whales are probably the most difficult animals of all to observe, since in the wild, many species swim at prodigious speeds and spend most of their time below the surface. It may never be possible, then, to infiltrate a school of free-swimming porpoises with a scuba diver or a pod of sperm whales with a sub.

As for studies done ashore, most species other than porpoises have never been kept in captivity because of their great size and food requirements. Among porpoises, the favorite research animal has been the bottle-nosed dolphin, like Flipper of television fame. Even here, though, there are problems. Dr. Cummings estimated that it costs about $25,000 a year just to keep a porpoise in a suitable tank and to provide adequate food and medicine. As for the larger cetaceans, Sea World once had a baby gray whale named Gigi. At the time she was released—because of the expense of keeping her—she was already twenty-five feet long, weighed eight thousand pounds, and was gulping down one thousand pounds of squid a day, which meant her food bill sometimes ran to more than $2000 a week. (As Dr. Cummings threw these figures at me, I was reminded of Paul Loiselle, who considered George Barlow's Midas cichlids too big to be convenient research animals. "Small is good," as he said.)

The current public interest in cetaceans—the widespread and controversial belief that they're uniquely intelligent among animals—dates largely from the publication of Dr. John Lilly's first book, *Man and Dolphin*, in 1961. Dr. Lilly suggested that because it has a large brain, the porpoise is probably extremely intelligent; that it may already have a language of its own and might also be able to learn a human language. Some of his animals did develop the ability to imitate roughly a few human sounds, repeating words and nonsense syllables in a Donald Duck sort of voice.

Dr. Lilly's theories, which were eagerly seized on by the general public, created a furor in the scientific community. Some researchers pointed out that brain size might not be directly related to intelligence and that, in any case, the porpoise may use its big brain mainly for echolocation. Others noted that parrots and mynas are even better at imitating than porpoises are, and that they're not unusually intelligent. Some believed it extremely unlikely that porpoises have anything that could be considered a language; others simply said that so far, there's no conclusive evidence either way.

Be that as it may, Lilly's theories did kick off a whole generation of ingenious experiments on porpoise communication. For example, in 1964, the late Dwight W. Batteau, a professor of mechanical engineering at Tufts University, invented an electronic device that could transmute human speech into whistles that sounded quite porpoiselike. Batteau then proceeded to put together a special vocabulary of simple words; some were English, some Hawaiian, and a few were made up, but all had lots of vowels, because consonant sounds don't whistle well. When spoken into the porpoise translator and broadcast underwater, these words emerged as sounds that could be distinguished from one another.

Next, working in conjunction with the Naval Undersea Research and Development Center, a predecessor to NOSC, Batteau trained two porpoises to carry out commands conveyed through the translator. In a design a little like Lana's computerized speech, every command began with the name of the animal it was directed to and then the word "imua," meaning "go ahead," and ended with the word "okay," the signal that the command was over and the animal was to do as it was told.

Unfortunately, there's no evidence that the porpoises responded with the kind of flexibility that one might expect if they themselves had a language similar to human languages. For example, if Batteau issued two orders instead of just one, in effect a compound sentence ("Maui, go-ahead, hit-ball-with-flipper, swim-through-hoop, okay"), the whistle for "swim through hoop" would trigger the ball-smacking response as the "okay" whistle usually did, and the porpoise would ignore the hoop altogether. However, Maui did learn to respond to fifteen different commands, uttered singly, in experiments conducted before Batteau's untimely death. So far, no one has tried to repeat or elaborate on this work, though there are at least three different whistled languages among humans that could conceivably be taught to porpoises. Words in these languages are created by shifts in pitch and power; similar shifts are to be found in porpoise signals.

In a different type of experiment in 1965, navy scientists Thomas G. Lang and H. A. P. Smith arranged for two porpoises to hold the equivalent of a telephone conversation. Doris and Dash (the latter worked for Batteau under the name of Maui) were put into separate tanks where they couldn't see one another. However, thanks to underwater mikes and hydrophones, they could—

whenever the equipment was turned on—hear one another, and they whistled and clicked back and forth quite enthusiastically. Dr. Cummings had a film of this experiment, and as I watched one of the animals sound off, then apparently listen for an answer before beginning to whistle again, it really did seem as if there were a conversation going on. I had to remind myself that other animals also take turns communicating: toadfishes do, for example, as do some insects, frogs, and birds.

The Lang and Smith experimental procedure lasted thirty-two minutes and was divided into sixteen time periods. In the even-numbered ones, the mikes were switched on, and in the odd-numbered ones, they were off; Lang and Smith recorded all the porpoise exchanges. Analyzing their tapes later, they found that they could distinguish six different types of whistle, and they labeled these A to F. Then they took Doris's tape and, in her absence, played it back to Dash. At first, his end of the new dialogue sounded much the same as it had in the original experiment: both times he tended to answer Doris's B-type whistles with Ds, for example. However, in period 8, he suddenly fell silent, and he maintained that silence until the tape was over. The following day the researchers played him the tape twice more, and each time he dropped out at about the same point, though now he resumed whistling in period 14. On closer analysis, it turned out that he became silent right after Doris's first F-type whistle and chimed in once more during the first period in which she produced no Fs: period 14.

Of all the sounds Doris and Dash made, the F whistles were the longest, the most complex, and the least stereotyped, so it stands to reason that they may also have been the most capable of carrying information. To Dash, something about that F whistle may have seemed not quite right. Perhaps it was the quality of the recording or the playback system that was was to blame, or perhaps at that point in the new dialogue, an F was simply a non sequitur.

Dr. Jarvis Bastian, a professor at the University of California at Davis, tackled the porpoise question from an entirely different angle. He presented the animals with a problem he thought they could solve only if they could communicate symbolically—in other words, if they had something like human language. Specifically, Doris, who was teamed this time with a colleague named Buzz, had to "tell" Buzz whether to push the paddle on the right or the paddle on the left. If he got the message, they were both rewarded with fish.

During the experiment, Doris and Buzz shared a circular tank that was divided down the middle by a heavy canvas curtain so that they could hear but not see one another. On Buzz's side of the tank were two paddles and one light, while Doris had two paddles and two lights. At the start of each trial, lights went on on both sides of the curtain to let the animals know that fish rewards were available. Then Doris's second light would come on to tell her which paddle to press: if it flashed, she was to nose the left-hand paddle; if it held steady, the right. However, the automated fish carousel would spill out her reward only if she could somehow get Buzz to press the correct paddle, also.

In the navy film Dr. Cummings showed me, Buzz's performance was stunning. Doris's light would go on, both would head for the paddles, and Buzz would hit the correct one about ninety percent of the time. Doris's vocalizations were, of course, recorded and painstakingly analyzed, and Dr. Bastian found that she generally produced trains of clicks when the light was steady and none when the light was flashing. Fourteen months later new trials were run with the code reversed, so that the flashing light now meant "Press the right-hand paddle" rather than "Press the left." Buzz was soon responding just as accurately, but this time analysis of the tapes showed that Doris was producing an entirely different set of vocal cues: when the light was steady, she would sound off sooner after it went on than she did when it flashed and would click faster and longer.

This sounds as if it might be safe to conclude that porpoises have, at the very least, a way to say "right" and "left" with clicks. However, another explanation is possible and more probable. Doris may not have intended to signal to Buzz at all. Instead, in the first experiment, for example, she may have been inadvertently conditioned to respond to the steady light by clicking—just as Pavlov's dog was conditioned to salivate when the bell sounded. Buzz in turn learned, perhaps fortuitously at first, that when he heard Doris clicking while his light was on, he would get fish if he pressed the right-hand paddle. It's the old problem again: how on earth does one determine what an animal's intentions are, assuming that one is willing to believe that animals are capable of having intentions in the first place? In this case, the fact that Doris clicked at the steady light even when Buzz wasn't there in the tank to hear her suggests that if she *were* communicating, it was accidental. At any rate, though the Bastian experiment didn't, after all, prove that porpoises have a language, neither did it prove that they don't.

Porpoises have also been studied, as so many other sound-producing animals have, by human observers trying to correlate particular sounds with particular behaviors. In general, porpoises make three quite different sorts of sounds: they click while echolocating and perhaps for communication as well; they produce doglike barks, squeaks, and squawks; and they whistle. It's the whistling that is most often mentioned when people speculate about a porpoise language. However, Dr. David K. and Melba C. Caldwell of the University of Florida, who have tried to match vocalizations to behavior, believe that porpoise whistles are fairly stereotyped and serve largely as the signature of the individual animal, to announce its presence, to identify it, and perhaps also to express emotions. For a cetacean, a signature feature might be very important, since visibility is limited underwater. Dr. Cummings cautioned that it takes a long time and a multitude of observations to prove that a particular sound is the signature of a given individual—that the sound is not instead used to signal distress, aggression, or whatever. All such possibilities must be taken into consideration.

Obviously we've still a long way to go before we can judge the level of porpoise communication; but in the absence of hard evidence, scientists have come up with some fascinating theories. For example, in one of the films Dr. Cummings showed me, there was an interview with Professor Kenneth Norris of the University of California at Santa Cruz. Norris speculated that one porpoise may be able to read another's emotional reactions by echolocating into its internal anatomy, including the air spaces there. Since the streamlined porpoise can't very well express emotions with its face or body, perhaps it has evolved so that its inner spaces change shape in significant ways under the impact of various emotions.

I was reminded of another possibility that Dr. Norris mentioned in his book, *The Porpoise Watcher*. He theorized that because porpoises don't use and have no need for tools, they may not share the human fondness for cause-and-effect sequences in language, and so they may have developed a system of communication quite unlike language. He wrote that " . . . our strange method of acoustic communication is almost grotesquely clumsy and difficult. What other creature, for example, would want to wait around for the message involved in a long sentence; to wait until all the little abstract symbols like prepositions and adverbs and participles were arranged according to an arcane plan before meaning could be extracted? Only, I suggest, an animal deeply involved in cause-and-effect sequences . . . " Norris believes that porpoise communication is probably about relationships and emotions and not about complicated abstractions.

Anthropologist Gregory Bateson also subscribes to this theory. Writing in the book *Mind in the Waters*, he pointed out that land mammals talk about the nuances of love, hate, and respect with posture and facial expression, with raised hackles, lowered tails, and so on; but the ocean has streamlined the porpoise, stripping away most of its ability to exchange signals by body language, and so perhaps it must use sounds, instead.

Whatever the significance of porpoise communication, the fact is that in recent years, porpoises—and the larger whales, too—have caught the public imagination. With some species in danger of extinction, we have belatedly begun to take an interest in the planet's largest and most mysterious mammals. As Dr. Cummings said, "The whale is an animal that has taken millions of years to evolve into something we don't even understand. The possibility that certain species may become extinct because of human shortsightedness makes one wonder about human intelligence."

One of the last films Dr. Cummings showed me dealt with some of his research on baleen whales. The species involved in this instance was the glorious humpback, a sleek charcoal-black giant with a graceful wingspan of long, slender white fins. The humpback is known for its singing—which is actually available on records—and Cummings and colleagues Paul Thompson and Steve Kennison flew to Hawaii to record its voice. They took along a sonabuoy, which floats on the water like a regular buoy but comes equipped with a hydrophone and radio antenna. Dropped in the bay at Maui with whales nearby, this listening device transmitted their songs to recording equipment several miles away. Simultaneously, recordings were also made from a ship close to the whales.

The humpback's song seems, at first hearing, endlessly varied, a free-form monologue of rumblings, howlings, mooing, and yipping. However, when Dr. Cummings brought his recordings home and ran them through a real-time sound analyzer like the one Peter Marler uses at Millbrook, he confirmed for the Pacific what he and East Coast researchers had reported for the Atlantic. The humpback's song is very like bird song: it's composed of stanzas (a phenomenon described for the first time by Cummings and co-worker L. Philippi) and it's repeated at intervals—though it may last as long as twenty minutes and then be recycled without a break. There were also indications that the song of a Pacific humpback is somewhat different from that of an East Coast humpback, which may mean that these whales—like some birds, frogs, and so on—have dialects.

After the films were over, Dr. Cummings took me back to his trailer office to see what humpback song looks like after it has been run through an analyzer and converted into what's called a waterfall display. The result was a long strip of paper covered with row upon row of what looked like tiny mountain ranges, with a definite three-dimensional effect. Those mountains showed both the way the whale's song changed in pitch over time and—with that suggestion of depth—the variations in loudness. However, this wasn't a detailed, second-by-second record; instead, the analyzer had been used to compress time, to get a broad-brush impression in which gross characteristics and patterns could stand out over long periods of time. And they did: every seventeen minutes, a very similar sequence of peaks reappeared—it was as regular as wallpaper.

This particular lab at NOSC also had the inevitable sonographic analyzer, used for producing voice prints, and Cummings's colleague Paul Thompson showed me a spectroscope as well, a machine that looked like a tiny television set. On the screen was a green grid and a dancing green line, like a live graph. As Thompson fed recorded humpback song into it, the spikes of the graph repeatedly tossed and fell, reflecting changes in pitch over time.

I also met Dr. Richard Johnson, who was in charge of a sound synthesizer—the whale people may actually soon be ahead of the bird-song researchers in this—and Cummings explained how it works. To synthesize whale sounds, scientists first record them and then turn their various characteristics into numbers, which can be fed to a computer. The computer produces graphs of wave forms, and the researchers use these to judge which are signals and what is extraneous noise. Then they ask the computer to yield just the signals. Using another method, Dr. Cummings had already synthesized some killer-whale sounds, and he planned to play these rather imperfect imitations to the animals he was working with at Sea World and in the wild, to see how they would react. Like other researchers studying animal sounds, he planned to tinker with the physical characteristics of the synthesized signals, to find out which elements carry the message.

Dr. Cummings and NOSC colleague Dr. James Fish have also used recordings of real killer-whale vocalizations to try to set up a kind of audible fence against whales that were competing with human fishermen. Thanks to Cummings's technique, developed in an experiment with the migrating gray whales, white whales can now be kept from invading Alaska's Kvichak River to prey on the annual run of young salmon. White whales, which are also called belugas, are small, at most eighteen feet long, and snow-white, with comical, friendly faces. They used to ride up the river on the flood tide twice a day during May and June, making huge inroads on the population of young salmon as they swarmed out to sea. Nowadays, apparently intimidated by underwater broadcasts of killer-whale screams, they mostly group at the mouth of the river, looking as if they'd like to go up it but don't dare.

"We got a lot of satisfaction out of the work in Alaska," Dr. Cummings told me. "People outraged at what the belugas were doing to the salmon run had been known to shoot them, and they're beautiful animals. When we found an effective way to keep them out of the river, who knows how many were saved? Of course, the method may actually be a hardship for the white whales, since it cuts down on their food supply—no study has been undertaken yet to find out how this affects them; but a hungry animal will work hard to find an alternative food source. And we mustn't forget that humans and their needs are also a natural part of the environment."

Dr. Cummings also used killer-whale playback in an unsuccessful attempt to protect pilot whales from angry squid fishermen who sometimes shot at them. Off the coast of southern California, boats go out at night and turn on lights close to the surface of the water to attract squid. So many come up in season, Cummings said, that you can almost walk on them, and fishermen net them or vacuum them up. However, pilot whales were regularly preying on the squid before they could be caught. Dr. Cummings tried bombarding them with killer-whale screams, but unfortunately they failed to retreat. He believes they may have been too excited about that banquet of squid to be scared off.

Scientists at NOSC are also hoping to use sounds broadcast underwater to help porpoises escape the nets of tuna fishermen. Tuna boats make their catch with a wall of netting about a half-mile long that runs down three hundred to four hundred feet into the water. The net is wrapped around a school of tuna, and the porpoises that often swim with them, and then the bottom is pulled together like a drawstring purse so that the fish can't escape. The porpoises seem to panic then and will race frantically from side to side; often they become caught in the net and drown. They should be able to jump out, but they generally don't try. One possible reason may be that noise from the boat interferes with their sonar. A solution might be to record whatever sounds they make when they do escape and play them back from outside the net to induce them to leap out. However, to do this successfully, scientists will have to know what noises they're competing against. Hence a team from NOSC, under the leadership of Dr. William E. Evans, was getting ready to go along on a tuna boat, to record and analyze all the noises of a fishing operation.

Dr. Cummings has also tried his killer-whale recordings on southern right whales. These creatures, now on the verge of extinction, are vast and relatively slow, though in their ponderous way they seem quite playful. They will often thrust up into the air and crash back into the sea again, or they'll slam their great tail flukes on the water over and over.

Back in 1971, Cummings and his associates played recorded killer-whale sounds to southern right whales gathered off the coast of Argentina. "We couldn't believe it," he said. "There was no obvious response." Instead of turning tail, as the gray whales had done, these animals continued whatever they were doing, though they would frequently spy-hop, raising their heads out of the water to look around. "We tried again and again," Dr. Cummings said. "The next year we came back with better and more powerful sound projectors and put models of killer-whale fins in the water while we broadcast, to help the illusion along, and there was still no apparent reaction."

It wasn't that southern right whales had no experience with the *orca*, since killer whales do hunt in those waters. In fact, Cummings and his associates once witnessed an attack. It happened just at dusk: he and his colleagues had been following a pair of right whales who were rolling near the surface, apparently courting, when a pack of five killer whales appeared and swam straight at them. "It was spectacular," Cummings recalled. "The right whales came together like one whale; they used their flippers and tails, rolling around in the water, apparently trying to strike the killer whales." The episode lasted twenty-five minutes, and afterward the right whales, who seemed uninjured, swam into shallow water and simply lay there as if exhausted.

Puzzling over the fact that, despite such incidents, the right whales definitely didn't find his recordings alarming, Dr. Cummings speculated that killer whales in those waters may have a southern dialect. His recordings were made in the northern hemisphere of the Pacific and are noticeably different from recordings he has since heard of Antarctic killer whales, However, as with porpoise signature whistles and humpback whale songs, it would take a lot of study and the process of elimination to demonstrate that dialects are involved, rather than just different items from the same species-wide vocabulary, expressed in different behavioral states.

Playback experiments such as the ones Bill Cummings has done are among the most valuable tools in the researcher's repertoire. They've been used, as we've seen, in studying crickets, birds, frogs, and other animals. The technique has even been tried by amateur naturalists at times. One man I know of used a commercial recording of wolves howling to get rid of an unusual garden pest: Dr. David Hellyer, a pediatrician, lives in the midst of a wildlife park—the land was his gift to the city of Tacoma. One of the other residents of the park was a moose named Chocolate, who was raised in a zoo. A few years back, during the rutting season, Chocolate took to hanging around the doctor's house, trampling the azaleas and looking as if at any moment he might charge the picture windows. Dr. Hellyer's attempts to drive him off by playing hard rock or classical music at top volume were unsuccessful; however, when he put the wolf record on his turntable, opened all the doors and windows, and turned up the sound, the moose immediately took off.

The first day that I spent with Dr. Cummings was mostly talk and films, questions and answers. On the second day, he took me to Sea World to meet Kona and Kilroy, the killer whales he was studying.

Killer whales, he told me, are the top predators in the sea and they are often likened to wolves. In addition to fish, they may dine on porpoises, sea lions, and even on other whales, while nothing dines on them. They're almost frighteningly intelligent: if they are chased by men in a boat, they will try different tactics to shake off pursuit, first disappearing under the water, then racing at top speed on the surface, then splitting up, with only one or two of the group remaining highly visible, apparently to draw off pursuit. Attacking a school of porpoises, a pack of *orcas* will encircle them, close in, and take turns picking off individual animals. Attacking a whale perhaps twice their size, they charge as a group. Once hated and feared by humans, they have come to be regarded quite differently since the first killer whale was captured in 1965 and taught to perform in an oceanarium. The whale, Namu, also starred in a Disney movie.

Bill Cummings is keenly interested in killer whales for three main reasons: because not much is known about them, because they're very vocal, and because they're not hard to find to study. In the fall of 1976, he was four months into his research. Together with his assistant, Marilyn E. Dahleim, he had recorded more than sixty half-hour killer-whale shows at the oceanarium (apparently *orcas* talk underwater to themselves or to their fellow actors while performing) and many more hours between shows. He still couldn't positively associate a single killer-whale sound with a particular behavior—"but we'll work it out," he said cheerfully.

Hiking across the vast grounds at Sea World with Bill, I passed tranquil ponds where brilliant pink flamingos grazed and tidal pools with starfish as brightly colored as gumdrops resting in their depths. Outside the theater where the killer-whale shows take place, there were porpoises in a pool, and as we passed, they swam alongside us, curious as puppies. They seemed even more like puppies once I'd seen Kona and Kilroy—huge by comparison and elegantly attired, much as penguins are, in black and white. They were eyeing each other across the barrier between their tanks and calling back and forth with sounds like the mewing of sea gulls. *Orcas*, Bill told me, are just as curious as porpoises, and they learn much faster.

He explained that so far he and Marilyn had broken down killer-whale sounds into twenty-five general types that they could both recognize. These included creaks like a squeaky door opening, bangings, tappings, tweets, and shrill, sea-gullish whines. There were also upscreams, which began high and rose still higher, and downscreams. All these signals were actually composed of clicks produced at machine-gun speed, so that to the human ear, they blurred into continuous sounds. The faster the clicking, the higher the pitch—a principle you can demonstrate for yourself by running a fingernail at various speeds across the teeth of a comb. Bill had also found evidence of whistles, relatively pure tones that were not composed of clicks.

The research was done, using a two-channel recorder, by taping the whale's sounds on one channel and human observations simultaneously on the other, so that on playback, one could listen to the whale signals together with comments on what the animals were doing at the time. During the summer, Kona was the only whale available to record, because Kilroy was away performing at Sea World in Ohio. He was brought back to San Diego in September, and for the first few days, he made very few sounds. Then he was moved into the tank next to Kona's, and suddenly there were all kinds of vocalizations to record. Kona would start a sound, and halfway through, she'd stop and Kilroy would finish it. She would make an upscream and Kilroy would imitate her, and then they'd both make the same sort of sound together.

Bill had been hoping to be able to tell Kona's vocalizations from Kilroy's but couldn't yet do it. In fact, when both whales were underwater, trying to decide who was sounding off was about as easy as trying to decide which of two ventriloquists was speaking for a dummy. Of course, if Kona had her head out of the water while Kilroy was under and the sound was audible in air, the answer was obvious.

Sooner or later, the behavioral significance of some of the *orca*'s sounds should become clear. The next step would be playback experiments, to see how Kilroy reacts to recordings of Kona, for example, or to synthesized killer-whale sounds. Eventually Bill plans to use a video camera or time-lapse photography to collect a vast quantity of data on behaviors and the sounds that go with them.

As Bill and I talked, Kilroy's trainer came by, and Kilroy immediately swam over to the side of the tank, fixing the young man with his inscrutable whale eyes. (There's no question that *orcas* can tell one human from another.) During the show, the trainer signals his whale either with hand gestures or with sounds, inaudible to the audience, broadcast underwater.

Soon after that, the whale show began. Kona was on that day, and she performed flawlessly. While she was submerged, I'd had no real sense of how big she is—she's just over seventeen feet long—but when she leaped, it was as if a Fifth Avenue bus had suddenly soared effortlessly into the air. The sense of power, of unimaginable strength, was breathtaking. When the show was over, Kona apparently still wanted more. She thrust herself up onto the stage, as she'd done at one point in the performance, and posed with tail flukes gracefully hoisted into the air. That wasn't part of the routine, and her trainer tried to entice her back into the water. He even gave her a push, but he might as well have tried to shove Mount Rushmore, and it wasn't until he finally disappeared backstage that she slid back into the tank.

Still we lingered on. I couldn't tear myself away, and it seemed Bill couldn't either, though he had seen the show fifty times or more. When most of the crowd had melted away, we had our reward: the gate to Kilroy's tank was opened and he swam out into the show tank. Evidently practicing a new trick, the two whales circled in opposite directions, one clockwise, the other counterclockwise, and then, on a signal, made straight for one another, took off simultaneously, and passed in midair. Nureyev and Fonteyn couldn't have timed it more precisely.

For me, though, the show was over; it was time to go home. As Bill and I headed back across the park, we detoured several times so that he could point out his favorite places, pausing at the dovecote, where one can handle live doves, and by the petting pool, where I tried once again—unsuccessfully—to make skin contact with a porpoise. Well fed, they drifted just out of reach, eyeing us tolerantly.

Sharing Bill's enthusiasms, I suddenly remembered a man I

met at the annual Animal Behavior Society meeting in 1976. At a party one evening, everyone else was standing around talking about how hard it has become to get tenure, postdoctoral fellowships, and even funding for animal research. But this particular postdoctoral student said with a gleam in his eye: "I can't complain. You know, it's still unbelievable to me that we actually get paid for doing what we do, when it's so much fun . . . "

14.

Canine Communication

He came out of a doze that was half nightmare, to see the red-hued she-wolf before him. She was not more than half a dozen feet away, sitting in the snow. The two dogs were whimpering and snarling at his feet, but she took no notice of them. She was looking at the man, and for some time he returned her look. There was nothing threatening about her. She looked at him merely with a great hunger. He was the food. Her mouth opened, the saliva drooled forth, and she licked her chops with the pleasure of anticipation.

—WHITE FANG,
by Jack London

I first met Michael Fox at a seminar he gave at Rockefeller University in New York back in 1975. His subject was wild canids—wolves, coyotes, foxes, and the like—and he had some fascinating things to say about canine communication. However, what I remember most clearly was the way he ended the lecture—with a strong plea for conservation. Since wolves are in danger of extinction, he said, scientists can't afford to simply sit back and look on, holding to their traditional objectivity. From comments I overheard as the audience of graduate students filed out the door, I gathered that such a plea from a research scientist is rare; but then Dr. Fox is an unorthodox sort of man, a veterinarian and psychologist, author of popular books as well as texts. In fact, by now he has moved on from research to become head of the Institute for the Study of Animal Problems, which is a new division of the Humane Society of the United States.

Over the years, Dr. Fox has observed wolf packs in the United States and wild dogs (called dholes), in India, and he has owned foxes, jackals, coyotes, and wolves, as well as a number of dogs. He has written about dog communication, drawing on both his own observations and on insights gained from studies of wild canids. It was primarily dogs that I wanted to talk to him about when I interviewed him, some months after his Rockefeller University lecture, in the Manhattan apartment of a mutual friend. It seemed to me that a book on animal communication should include something about the animals who live most closely with humans.

However, it turned out that we may know more about the signals of birds and bees than we do about those of dogs, for dogs haven't been studied as much. Perhaps *Canis familiaris* doesn't seem a "real" canid to scientists, as a canary didn't seem a "real" bird to Donald Kroodsma. Animals evolve to fit their natural habitat, and when humans step in and domesticate them, breeding them for particular purposes, they corrupt the delicate and fascinating relationship between animal and environment. By studying pigs or cattle, dogs or canaries, we might learn what humans want of such species, but not what evolution has demanded.

Studies of wolves and coyotes, however, have been forging ahead. Realizing that wolves are a threatened species, scientists have rushed to observe them while there's still time. What makes them particularly fascinating is the fact that in some ways they're similar to humans: they're hunters who live in packs that are rather like extended families; they're intelligent; they share food and cooperate in the business of survival; they teach their young and display deep affection for one another. And if wild canids are fascinating in themselves, knowing something about them is also a help in understanding dogs, for the dog is not only related to the wolf but is believed to have descended from it or to share a common ancestor.

Both wolves and dogs communicate primarily in three ways: with pheromones, with sounds, and with body language. In his book *Understanding Your Dog*, Dr. Fox pointed out that one of the things that's so appealing about dogs is that their emotions and reactions are easy to read: they seem straighforward, honest, utterly incapable of deception. Dogs have no trouble reading people, either, and in fact they may be better at interpreting us than we are them. Dr. Fox also said that wolves he has had have been even more expressive and more sensitive to human nuances than most dogs are.

There are several reasons why humans and dogs can communicate so comfortably. One that's often overlooked is the fact that individual dogs train their owners to respond to particular signals. For example, one dog I know notifies her people that she wants to go out by popping up and down like a kid on a pogo stick in front of the door. (Cats do the same sort of thing. One of mine uses the classic Lassie technique to tell me she's hungry: she runs ahead of me toward the kitchen, looking back to make sure I'm following. My other cat conveys the same information with an even more effective signal—when her dinner is late, she steps in front of me and trips me.)

However, the most important reason that dogs are easy to read is that their body language is in many ways similar to ours. This is particularly true of facial expressions. The delighted grin a dog wears when greeting its owner or a canine friend looks very much like a human smile. The dog's play face—which also involves a grin but with a wide-open mouth and ears pricked forward—resembles human laughter, and though humans rarely produce a full-fledged, teeth-baring snarl, the impulse is there, and it helps make the threat in a dog's snarl unmistakable.

Canid postures are also hauntingly familiar. Dr. Fox has photographs of wolves, for example, in which it's easy to tell the pack leader from his subordinates. The alpha male stands tall and holds his head high and his ears forward. The others seem hunched and wary—ears back, tails down, obviously avoiding their leader's eye. In some cases, heads are down as well, just as humans bow their heads in submission. Canids out to dominate or threaten make themselves appear larger and more dangerous by raising their hackles and walking stiff-legged—they seem almost to stand on tiptoe. To signal submission, they make themselves smaller. Perhaps this is simply the logic of the body, since many animals, from wasps to humans, share the same tendencies.

In canid society, a direct stare is a threat, and again this is a signal many species share. Dr. Fox explained that in a wolf pack, the dominant animals can control the others from a considerable distance with eye contact. All they have to do is stare intently at a subordinate, and it will generally assume a submissive posture. Ethologist Konrad Lorenz, observing that a threatened animal will often turn its head sharply aside, suggested that it is offering its throat to the aggressor, which would be, perhaps, the ultimate in conciliatory gestures. However, Dr. Fox believes that all that's involved is an exaggerated looking away. In any case, he recommends that when humans are threatened by a strange dog, they should definitely not return stare for stare.

Canids are, of course, intelligent animals, and so their signals are flexible and sometimes even subtle. For example, though looking away is usually a sign of submission, on rare occasions it can be a bid for dominance instead. Dr. Fox recalled a particular wolf who was once introduced to a captive pack. On previous occasions, newcomers had been attacked by the pack in spite of the fact that they assumed submissive postures and facial expressions. In this case, however, the stranger refused to cringe and instead simply ignored the other animals, staring off into the distance as if they didn't exist. Surprisingly, the attack was aborted. Dr. Fox explained that it can actually be more intimidating to ignore threats and look away. "It's like a man wearing sunglasses or a mustache," he said. (He was at the time sporting quite a luxuriant mustache himself.) "The lack of expression creates distance, and that distance creates social control and dominance."

I asked Fox how you can tell, when you're out walking a dog and a strange animal approaches making threatening noises, whether a real fight is likely to develop. He said that animals who are serious about fighting don't generally make a lot of noise or display very much—snarling with hackles raised, for example. Most often, when dogs threaten, all they do is stand over the other and shove: the aggressor dog will place itself perpendicular to the other animal, shoulder to shoulder, and push. In fairly short order, one individual or the other gives up, rolls over, and starts whining. When this kind of ritualized wrestling match is all that's involved, it's often best to just let it happen, since by intervening, you might start a real fight. In fact, on a quiet street, you should probably drop the leash and move away so that your dog won't take it into its head to defend you as part of its mobile territory. If you're really afraid a serious fight might break out, you can always ostentatiously pick up a rock, even an imaginary one. Most dogs will get the point.

Body signals to beware of, on the other hand, are those typical of the fear-biter: the animal puts its ears back and tucks its tail between its legs as if in submission, while at the same time, the hackles are up and the dog is growling. It may actually grin—and simultaneously expose its teeth in threat. Believe the growl, not the grin, Fox said, and steer clear of the animal. A wagging tail is also not always what it appears to be. Though it's usually a token of friendliness, a tail carried high and waved stiffly and rapidly often accompanies other signs of aggression.

Some of the signals in the dog's repertoire are, in effect, baby talk, infantile behaviors carried over into adult social relationships. For example, for the first few weeks of life, a dog can't empty its bladder or bowels without help from its mother. When the mother returns to her litter, she will nose the puppy's groin. It responds by holding perfectly still so that she can lick the inguinal area and thus trigger evacuation. In general, she then tidily laps up the urine or feces. In adulthood, many dogs will still become momentarily immobile if touched on the groin, and a dog that has been scolded by her human will often raise one forepaw, not to shake hands, but in the old wolf-pack signal that she intends to roll over. The she may roll onto one side, exposing her genitals just a a puppy does for her mother. She may even urinate a bit as she did when she was small. This is a sign of extreme deference, and in such a situation, a dog who piddles at her owner's feet is really presenting a bouquet, according to Dr. Fox. (Similar occurrences take place in wolf packs between adult wolves, and there the dominant animal may actually lick up the urine afterward, just as a parent would.) When a dog directs such signals to a human, it's because it relates to its owner as if he were the pack leader. In fact, a dog's human family *is* its pack.

Among wolves, there are other signals as well that linger after babyhood. Once the cubs are a few weeks old, the adult wolves begin to bring food home to them from the hunt. Since they're not supplied with pockets, they carry it in their stomachs and regurgitate it when the cubs lick them about the face and mouth. (Mother dogs also sometimes regurgitate to their litters.) Low-ranking adult wolves often use this cub gesture as a sign of submission, poking at the muzzle of a dominant animal, behaving as if they were cubs and the other animal were the parent.

Just as canids use baby signals in adult situations, so they sometimes use aggressive displays to express ritualized allegiance and affection, or so Dr. Fox believes. He once watched while several wolves came up to their pack leader and made submissive gestures. The alpha male got up in leisurely fashion and pinned them to the ground, one at a time; then he gave a big yawn and sat down again. Dr. Fox believes that this was a case where aggressive signals functioned instead as bonding behavior, and he thinks this may often happen.

Obviously canid body language is versatile and complex, and the more sociable the species, according to Dr. Fox, the richer the signal system. Foxes, which are fairly solitary animals, seem to have no submissive behavior—most of the time, they simply avoid one another—and they also lack subtle gradations of intensity in their facial and body displays. Wolves, within the complex social structure of the pack, have the most impressive signal repertoire of all.

Remembering Roger Fouts, who believes he communicates with his chimps in a kind of pidgin language, part signing and part Chimpanzeese, I asked Dr. Fox whether he ever communicated with canids in their own terms. He immediately recalled the one and only time he was ever attacked by a wolf. He went into the animal's cage, he said, not realizing that its female was in heat. Inadvertently he came between the two—and suddenly found himself pinned against the cage with the huge male standing over him. "I looked away and whined," he recalled. The wolf stopped attacking and someone else pulled him away.

Dr. Fox often communicates with dogs in a dog way. "It sometimes freaks them out," he said. "I use their body postures, facial expressions, vocalizations. I paw at them, touch them on the groin, bite them on the muzzle, make the play-inviting bow wearing the play face. Some of them bristle, yelp, run away, get very disturbed—for them, it's a schizoid space. Others get right into it. Though I'm sorry when they're disturbed, it's fun when they play along."

Though body language is important to dogs, they also rely on pheromone communication, to what extent we really don't know. It may be impossible for humans to understand or even imagine the olfactory world of a dog—a world in which odors linger on for hours to tell their stories.

Dogs come well equipped for sending odor messages. Each has a special scent gland near its tail—which explains the avid canine interest in that area—and there are other glands between the toes. Dr. Fox explained that when a dog urinates and then scrapes at the spot with its paws, it's marking the site visually, but it's also adding to the scent message from those interdigital glands.

With their system of marking with urine, canids can communicate across time. The spaniel lifting his leg beside a tree trunk is leaving a calling card that probably conveys several different messages. It may be a territorial marker or an aid to finding his way home; probably it also identifies him to other dogs who know him, and it may even specify his sex and tell them how long ago he was there. But we haven't really broken the code of canine urine yet—we can only guess what the messages are.

I asked Dr. Fox why, when you're walking two dogs, one sometimes insists on sniffing interminably around a tree or bush while the other remains indifferent. He explained that the animal more interested in sniffing and marking may be the more dominant of the two, since these activities seem to be more important to a dominant animal. Then again, the difference might be simply one of gender: male dogs are usually more interested in sending and receiving odor messages than females are—though females in heat do engage in a lot of marking.

Dogs also signal, of course, with a variety of sounds, and to a surprising extent, they can understand human speech. Dr. Fox estimates that many dogs can understand forty words or so; sixty are possible but would be exceptional.

Dr. Norman Bleicher, now at International College in Los Angeles, recently completed a study of dog vocalizations. Using the classic method, he observed behavior as he recorded sounds to be analyzed later with a spectrograph, and he found that there were indeed sounds that seemed to go with particular situations. Puppies, for example, vocalize differently according to whether they're hot, cold, hungry, hurt, playing, and so on. Dr. Bleicher also noted that dogs bark and growl both in play and to threaten, and that although some dog owners claim they can hear the difference, it doesn't show up on a spectrogram. This could be because the human nervous system is in some ways capable of making finer distinctions than the spectrograph can, or it could be, as Bleicher suggests, that barks in particular may serve merely to announce that the dog is there, while the animal signals its intentions with body language. At any rate, the bark is usually a long-distance threat, and so when two dogs who are face to face bark at one another, it's safe to assume they're playing—unless theirs was a sudden and unexpected confrontation.

Scientists are also trying to decipher the vocalizations of wild canids, and the great mystery there is group howling. Dr. Erich Klinghammer has been studying and recording the sounds of a captive wolf pack in Indiana, and at the Animal Behavior Society meeting in 1976, he warned me, "Don't let anyone tell you we know what the wolf chorus is about, because we don't.

"Solo howls are not difficult to figure out," he said. "They apparently serve to attract a mate in the breeding season and to make contact with other wolves at any time. Probably the wolves can recognize one another's voices, just as I can, and in fact group howls may be one way wolves learn the voices of other pack members. They may also serve in territorial defense, to warn off other packs, as Harrington and Mech suggest, although this has not really been proven to my satisfaction. When humans howl in wolf territories, they rarely come to investigate, so we do not really know how wolves see this."

Why do so many people have such strong feelings about the wolf, pro and con—even city people who have never really seen one? I suspect that there are those in both camps who tend to project human values onto the animal. Some see the wolf as a wild creature that is intelligent, affectionate, courageous, self-sufficient. Others see a cunning and dangerous animal who kills without mercy. The wolf seems to embody both our negative and positive feelings about our own role as predators.

Conservationist that he is, Dr. Fox is sometimes turned off by the way the wolf's case is argued. "I'm getting very annoyed with people who feel they have to justify the existence of wolves or whales or whatever by demonstrating their superior intelligence or sociability, in other words, their similarity to us humans," he told me. "Everything is precious—whether it's humanlike or not. What right have we to destroy *any* animal? That's why it's important to realize that animals communicate and feel much as we do. Perhaps out of that, we can develop what we really need: a biospiritual ethic, a reverence for all life."

15.

The Native "Language" of Chimps

It was only many days later that Ransom discovered how to deal with these sudden losses of confidence. They arose when the rationality of the hross *tempted you to think of it as a man. Then it became abominable—a man seven feet high, with a snaky body, covered, face and all, with thick black animal hair, and whiskered like a cat. But starting from the other end you had an animal with everything an animal ought to have—glossy coat, liquid eye, sweet breath and whitest teeth—and added to all these, as though Paradise had never been lost and earliest dreams were true, the charm of speech and reason. Nothing could be more disgusting than the one impression; nothing more delightful than the other. It all depended on the point of view.*

—OUT OF THE SILENT PLANET,
by C. S. Lewis

From the beginning, the enigma of chimps with language haunted me. Why have they been able to learn it so readily? And since they do have this unexpected aptitude, why haven't they developed some form of language of their own? Or have they?

Obviously the answer to this last question lies in studies of the way chimpanzees normally communicate. Though they are known to use gestures such as the palm-up begging gesture to signal to one another and, in captivity, to humans, and though they hug, kiss, and clasp hands in greeting and sometimes chuck one another under the chin—all actions that bear an uncanny resemblance to things humans do—scientists who have observed chimps in the wild have seen nothing like the elaborate, intricate, and formal sign languages that humans have devised. Yet one set of experiments done with captive chimps suggests that, despite this, they can indeed exchange a great deal of information—and not just about their emotions and intentions, but also about the world around them. In fact, Emil Menzel, the man behind the experiments, has suggested that in teaching chimps language, we may not be improving at all on their natural ability to communicate with one another. Premack and others have made a similar point: the main thing they've taught chimps is a system for conveying what they already know to humans.

Professor Menzel is now at the State University of New York at Stony Brook, Long Island, and on a spring day in 1977, I interviewed him there. A dark-haired, slow-spoken man of medium height with a quiet, dry sense of humor, he had recently acquired a pair of chimpanzee children caught in Africa just months before. Wild-born chimps, he told me, are in most ways more interesting than those born in captivity. Life in a cage or even in a captive colony somewhat distorts the chimpanzee personality. Nevertheless, since chimps in the wild are a dwindling species, Menzel hesitated a long time before ordering the animals.

In a long, low building that also houses birds, fish, and a colony of marmosets, the two chimpanzee children had been assigned a suite of rooms containing three connecting cages, the whole complex about fifty feet long. Their names were Maynard and Madeline, and when we entered their rooms, they were sitting on a high shelf in the home cage behind a wall of wire mesh. Madeline immediately bristled and began to make a deafening noise by banging on the shelf. Clearly she was making threats, though Menzel assured me that she was all bluff. The animals were about three and a half, the same age as Nim (see Chapter 4)—who was, incidentally, scheduled to pay them a visit in the near future. Unlike Nim, they were quite independent, attached to one another rather than to human beings.

While I watched, Professor Menzel prepared to open the connecting doors between the cages to give the young chimps the run of the whole suite. They obviously knew that their daily treat was coming up and they swung down to the floor, where they waited, visibly excited and touching one another for reassurance. The minute Menzel opened the home-cage door, they hurtled through it.

For the next half-hour, Professor Menzel quietly tape recorded a description of everything the young animals did as they tumbled from room to room. It was at times a rather breathless description, for they hardly ever stopped moving. They swarmed up and down the wire mesh, went hand over hand up ropes that dangled from the ceiling, scampered along a ladder that served as an overhead catwalk, played tag in three dimensions.

A number of times Madeline flung herself down inside a wooden crate to stare, fascinated, at her own image reflected in a small metal mirror that was fixed to one side of the crate. This brief respite usually ended when Maynard came running up to give the crate or her shoulder a thump; then he would back off, obviously daring her to chase him. Like a shot, she would leap out of the box, and the two of them would streak into the next room to play tag around the base of a tall wire cylinder, using it as if it were a tree trunk in the forest—feinting, dodging, reversing directions, dashing around and around it. When a timer finally signaled the end of the play period and Menzel shooed the animals back into their home cage, they plopped down as if they were glad to rest.

Joining me again in front of the home cage, Professor Menzel explained that he was interested in the cognitive aspects of play. Originally he had planned to spend just five to ten days observing chimp play periods, but they proved so interesting that he was now on his forty-sixth day. Over the period, the animals' activities had become more and more complex and elaborate, and increasingly they played together rather than separately.

Menzel's observations have picked up some fascinating clues to the way their minds work. For example, one day a radiator cover in the middle cage came loose. The chimps, cautious but obviously excited, immediately checked the cover on the other radiator in the room and then headed back into the home cage to inspect the radiator in there. Obviously they recognized that each was the same sort of object, but it was also clear that they carried around in their heads a mental picture—a cognitive map, as psychologists call it—of the home cage, its contents, and their location, since from the middle cage, they couldn't see the home-cage radiator. Cognitive maps interest Emil Menzel very much.

"The question is," he said, "what do animals know about the world they live in, and how do they get where they're going?" Some species, such as army ants, follow odor trails, while others can memorize a particular route. However, an animal that can carry in its head a kind of mental map of familiar terrain, that can remember where dozens of objects are located in it, is obviously doing something of a different order.

What is a chimp's mental map actually like, and could one learn to use a real map? In a study designed to explore these questions, Menzel had recently collaborated with David Premack and a colleague, Guy Wodruff, at the University of Pennsylvania. They presented four infant chimps with a "map" of an enclosure in the form of a closed-circuit television image. First a familiar caretaker, carrying the morning ration of fruit, showed himself to the animals where they waited in a blind at the edge of the enclosure. Then he walked around to the front of the blind until he reappeared on the TV screen, which enabled the chimps to watch him as he went out into the field and hid. Afterward, they were allowed to go looking for him.

In control trials, the procedure was much the same, except that the animals could watch through a window in the front of the blind as the caretaker hid. Though the chimps struck out with more confidence on the control trials and found the caretaker more easily, three out of four were also able to use the information from the television screen. Of course, a television image, indistinct though it is, isn't all that far removed from three-dimensional reality; the next step will be to test chimps with images more abstract and maplike, to see how far they can go.

In a long conversation that began over lunch and continued afterward in his office, Professor Menzel recalled his earlier studies of chimp communication. At the time he was working at the Delta Regional Primate Research Center in Louisiana, with a group of eight young wild-born chimps, four males and four females. The animals had the run of an enclosure almost an acre in size, overlooked by an observation house.

In the beginning, because they were very young, the chimps

were virtually inseparable. None of them was about to venture forth into the enclosure without a friend to cling to, and so they traveled around in pairs or as a whole pack. Often they resorted to tandem walking, proceeding Indian file with one directly behind the other, holding on at the waist. It wasn't the most efficient way to get around, because sometimes they got their legs tangled, and if they were upset or excited, there was a lot of stumbling, whimpering, and shifting as each animal tried to be the one to bring up the rear.

Gradually, though, they got the logistics sorted out, and soon they were traveling around the enclosure in surprisingly purposeful and coordinated groups. In fact, their apparent ability to decide instantly among themselves which way to go next was so impressive, Menzel said, that he almost began to believe they had a "group mind." He could seldom see any obvious signals passing between them—perhaps one time in ten there was a vocalization or gesture—nor was any one animal always the leader. Either they had an uncanny way of spotting the same goal—the same toy or piece of food or whatever—at the same instant, so that they moved out together, or they were signaling to one another in ways he hadn't been able to pick up. Betting on the second explanation, he decided to look more closely at chimp communication.

A scientist who wants to study animal communication can choose any of several different approaches. He or she can take a young animal and try to teach it human communication, as the chimp language projects have done, to find out what it's capable of. He can observe the animal and try to compile a kind of dictionary of the signals it makes, the vocalizations or body movements, recording the circumstances in which they occur, as Bill Cummings has been doing with killer whales. Or the experimenter can take an obvious example of coordinated group action and work backward from that to see how it comes about and what signals are involved. This is what von Frisch did in his studies of bees, and Emil Menzel's experimental design was strikingly similar.

In a series of experiments that went on, with many ingenious variations, for several years, his basic procedure was this: first the chimps were all locked up in a cage, from which they couldn't see into the enclosure. Then an experimenter hid food or another object somewhere in the field—stashing a pile of apples away in the tall grass, for example, or burying a bunch of bananas under a cover of leaves. Next, one of the chimps was removed from the cage, carried out into the enclosure, and shown the food; then the animal was returned to the cage. A few minutes later, after the experimenter had left the field, the whole group was released.

The chimp in the know, much as she might have liked to keep all the food for herself, was too timid to go after it alone, so she had to recruit a following. Though there were occasions when the other animals were uncooperative, most of the time some or all of them went along eagerly. The animals would emerge from the holding cage, the leader would take a few steps, look back to see that the others were coming, and then take off down the field with the rest close behind. The leader seldom had any trouble remembering where the treats were hidden. In fact, in some trials, as many as eighteen different caches of food were hidden around the enclosure, and not only did the chimps rememember where most of them were but—in an impressive demonstration of their ability to use a mental map—they created their own itineraries, going from pile to pile by routes much more efficient than the ones the experimenter took while sharing his secrets.

As the experiments became more elaborate, the chimps proved that they could communicate information beyond the simple fact that there was something desirable out there in the enclosure. For example, sometimes Professor Menzel hid a dead snake rather than a pile of food and then showed it to a single chimp in the usual way. Afterward, both leader and followers came out of the holding cage with hair on end and, instead of racing down the field as they did on food trials, they moved slowly and cautiously. When they neared the place where the snake had been hidden, they began to throw sticks and to slap the ground. Since the snake had, in the meanwhile, been removed from the field, they weren't being cautious because they sensed its presence.

Furthermore, the leader was obviously also able to communicate something about the relative quality of a food find. In a series of tests, two of the animals, Belle and Bandit, were taken out into the field one at a time; Bandit was shown a large pile of food and Belle a small pile, or vice versa. Later, when they were released into the enclosure together, they would set off in tandem, and over 80 percent of the time, they'd head for the larger pile first. Sometimes, instead, both were shown small piles or both saw large ones; this was done to make sure they couldn't solve the problem by a simple process of elimination ("If I saw a small pile, then she must have seen the big one . . .").

In an elaboration of the basic procedure, the whole pack was turned loose after two leaders had been shown piles of different sizes. Usually the group would split up and, again, almost 80 percent of the time the leader who had been shown the larger pile attracted a larger following. When this didn't happen, the explanation seemed to be not that there had been miscommunication, but that the animals consistently preferred to tag along with the leader who was more likely to share the food.

How did the chimps manage all this? Were signals exchanged back in the holding cage? As far as Menzel could tell, they weren't, except perhaps when it was a dead snake that was hidden, since then the leader chimp reentered the cage obviously upset and fearful. On food trials, the experimenters couldn't spot anything that looked like signaling in the holding cage, and even after the animals were released, there were seldom gestures as obvious as pointing or beckoning.

There are tremendous advantages to studying an animal as nearly human as the chimpanzee. One advantage became obvious when scientists, watching films made of the animals emerging from the cage, found that they themselves could generally predict within the first few seconds, just from the chimps' behavior, where the food was located, which animals knew the direction, and whether there was a smaller, secondary food pile somewhere.

The signals they picked up were subtle but simple: mainly they judged by where the animals pointed their bodies and their eyes. To find out whether the chimps were also cued in the same way, the scientists used themselves as leaders within the experimental format. For example, they would remove a chimp from the cage, and one human would hold him while another walked a few paces in the direction of the hidden food, glanced back, and then leaned forward as if he'd seen something. The chimp was then returned to the cage. When the whole group was released, he had no trouble leading them to the food.

Two other signals, though they were seldom used by the chimps themselves, worked equally well: the experimenter would stand beside the animal, facing in the right direction, and point; or he would walk a few steps and then point. If, however, he simply turned slowly in a circle, the animals searched at random or not at all. Similarly, the experimenters demonstrated by playing leader that when two piles of food were hidden, the chimps tended to follow the animal who started out fastest and went farthest before stopping to see whether he had a following. Apparently eagerness was the cue.

However, the animals also used signals that were even subtler. Menzel speaks of a "language of the eyes," something Belle in particular was good at. "A few times I carried her out into the enclosure," he recalled. "She'd watch me, but I refused to give her any clue, so pretty soon she would start scanning the field. When she was looking in the right direction, I'd pat her on the back and say, 'That's it,' and put her back in the cage. When we turned her loose, she'd head out in the direction of the food. First, though, she had to ask the question, as it were, with her eyes. But she was phenomenal. I don't know whether she was smarter than the rest or just a bigger chow hound."

Belle also supplied Menzel with evidence that his chimps weren't simply signaling one another unintentionally, for in a series of confrontations with another animal, she did her best to lie. This came about when, in some of the two-leader trials, she was paired with Rock, the dominant male in the group. No matter what size pile she was shown, Belle nearly always attracted the larger following because she was willing to share her find and Rock wasn't. After a while, however, Rock began to follow her and steal her food. Soon she stopped uncovering it and would sit on it instead until he went away.

A few days later he began to shove her aside and search the place she had been sitting. It took a month or so, but finally she learned to take up a position some distance from the pile for as long as he was watching; gradually she sat farther and farther away, and she would even look in a different direction. However, Rock still triumphed, because she always sat facing the food. He would search in that direction, watching her intently, and as soon as he got close to the goal, Belle would show signs of nervousness, so he'd concentrate on that area. Though she never did learn to hide her anxiety, a few times Belle successfully tried an entirely different strategem: she headed in the opposite direction from the food, and then as soon as Rock began to search, she doubled back quickly. The contest between Belle and Rock continued for months, and the steady escalation of tactics might have gone on indefinitely if Rock hadn't become sick.

Over the course of Menzel's experiments, the chimps became more and more subtle in their signaling. In the early days, quite obvious cues were sometimes used: a leader trying to recruit a following would whimper, beckon, tap friends on the shoulder, and so on. These signals soon virtually disappeared. Reporting on his experiments, Menzel wrote, "The stark fact . . . [is that] in the last months of research, by which time the chimpanzees were adolescent and showing their best performance in getting to the best available objects, we almost never saw these impressive displays. Indeed, even in the early tests it was Bandit, the most infantile and variable leader, who did the best and most frequent displays; and he showed them on a fraction of his successful trials. . . . I would guess that young and inexperienced chimpanzees have a *richer* 'vocabulary' of humanoid-like signals than older and more experienced ones and that they are reinforced . . . into abandoning them for something different."

As an example of the way the chimps' signals evolved, Menzel described the cues used to initiate tandem walking. In the beginning, an animal would simply grab a friend from behind, perhaps actually pulling her or him up to a standing position. However, within just a few months, the chimps had developed a whole set of less obvious signals. Now a tap on the shoulder, a begging gesture, a whimper, or a pout were all effective; sometimes a chimp would simply glance at a companion, then look out into the field, then present her back so that he could latch on. Obviously the animals were capable of agreeing on a "meaning" for certain signals—which is surely a prerequisite for the development of language.

Menzel's experiments, coupled with observations made in the wild, strongly suggest that chimps don't have a formal gestural language like the human sign languages, and he believes this is because they already have a means of communication that is, for their own purposes, just as good. In fact, playing devil's advocate, he has suggested that in teaching chimps language, we may not be improving at all on their natural ability to communicate or to conceptualize. He noted that, to date, the only evidence to the contrary is an experiment done with Lana. She was presented with two items, one she could see but not touch and one she could touch but not see, and was asked to say through her keyboard whether they were identical. All the items used were familiar foods: she might be shown a banana, for example, and then be required to reach through a portal and handle an apple, after which she was to type out either "same" or "no same." For whatever reason, she was significantly more accurate when the foods presented were ones she had names for than when they were equally familiar items for which she had never been taught a name.

Of course, future tests may confirm that it's an advantage to be able to think in words or word-symbols, but in the meanwhile, Menzel's point of view seems a useful corrective to the assumptions most of us make because we're unable even to imagine what it would be like to think without language.

Menzel feels that it's time to question and perhaps reject many of the old assumptions about animal communication: that animal signaling is mostly unintentional and involuntary; that without language, animals can convey nothing about the past or future or about objects not present; that an animal's ability to communicate is determined by the size of its "vocabulary"; that particular signals can be said to have particular, invariable meanings. The meaning of any signal, he explained, depends on the individuals involved and on the situation—the whole rich context must be taken into account. An incident that occurred during a visit he made to the Gombe Stream area in Tanzania, where Jane Goodall was studying chimps in the wild, illustrates his point quite neatly.

It seems that, together with a few of Goodall's students, Menzel was observing a chimp family—mother, infant, and six-year-old daughter. Eventually the mother, carrying the infant, got up to move on. The daughter stayed put. The mother hung around for a while, waiting patiently, then finally she reached out and picked a grass stem. Immediately the daughter came running, plucked a piece of grass herself, and the whole family set off down a path. The students explained to Menzel that they were undoubtedly going termite fishing, since the path led to a termite mound and chimps use grass stems to probe such mounds and pull out termites for eating. Menzel believes it's very likely that when the daughter saw her mother pick grass and start down that particular trail, she, like the students, realized what it meant. The gesture may even have been intended to communicate. Nevertheless, to decipher it, the chimp child had to know both her mother's habits and the terrain.

Still, the basic question persists: why do chimps have the ability to develop the rudiments (or perhaps more than the rudiments) of symbolic, syntactical communication when they apparently don't need and don't use that ability in nature? The question seems related to another statement I've read: chimps seem to have more intelligence than they really need to survive in the wild.

When I repeated that last statement to Professor Menzel, it reminded him of the old Darwin-Wallace controversy. Independently of Darwin and at about the same time, Alfred Russel Wallace also came up with the theory of natural selection. However, Wallace believed that in the case of humans, divine intervention must have directed and speeded up the process of evolution. As proof, he pointed out that people in primitive cultures had far more intelligence than their life-style demanded, though according to natural-selection theory, a species would develop a particular characteristic only if there were pressure on it to do so. "Natural selection could only have endowed the savage with a brain a little superior to that of an ape," Wallace wrote, "whereas he actually possesses one but little inferior to that of the average member of our learned societies." Therefore, it seemed to him that humans had not evolved in quite the same way as other animals.

Wallace's "evidence" was, however, not accepted by everyone, since many Victorians saw primitive peoples as a kind of missing link between apes and humans, as beings with reduced intelligence. There were even serious debates over whether certain tribes were actually human. Nor were Europeans the only people to question whether other races were genuinely fellow members of the species. Menzel recalled, "I've read a nice account by a Chinese who examined Marco Polo, questioning whether he was human, noting that though he was hairy and animal-like, he did have a form of speech and was capable of learning Chinese."

Today, of course, it's generally accepted that "primitive" peoples have as much intelligence as those who live in industrialized societies, and that no culture or language can be considered superior to any other. Each must be taken on its own terms. Menzel suggested that in the same way that we've learned not to underestimate other humans because their life-styles appear simpler than ours, we may have to learn not to underestimate apes and perhaps other animals, as well.

It also seems likely that we have misjudged the demands of life in the wild, for when ethologist Geza Teleki, who spent several years observing the chimps at Gombe, tried to learn termite fishing by watching the animals do it, he found that it took a surprising amount of skill and practice. He not only had to learn where the mounds were located but which particular tunnels on each mound were likely to prove productive—which probably required the chimps to memorize a hundred or so fishing sites. The twig or grass stem he used had to be just the right length and thickness and neither too flexible nor too stiff. It had to be inserted carefully, navigating twists in the tunnel, to just the right depth, and twiddled for the proper length of time—long enough to attract termites but not so long that they could simply bite off a piece. And finally the probe had to be withdrawn smoothly and at the right speed so as not to scrape off the insects, once again using wrist action to cope with twists in the tunnel. Teleki was never able to match the fishing prowess of some of the adult chimps he observed.

However, when he looked at termiting as a "subsistence technology" practiced by several different primates and compared the techniques of baboons, chimps, and humans (who also consider termites a delicacy), he found that there were indeed significant differences. The baboons used hands and mouth to collect crawling and flying insects as they emerged from the mound. Though they often watched the chimps fishing with twigs, they never tried it themselves. Humans, at the other extreme, used a wide variety of techniques, though their most interesting method was to imitate the sound of rainfall—by pounding on the mound with sticks, for example, or clicking with the tongue—because termites tend to emerge during a rainstorm.

For baboons, the termiting season is quite short. Chimps' fishing techniques prolong it considerably, and human methods are a further substantial improvement. However, Teleki's point was that there is no vast technological gap between the species—there are only differences in degree.

In the last few years, humans have begun to realize that apes are much more like us than we had thought. Recent biochemical studies lend a new credibility to this change in perspective. A comparison of the molecular composition of proteins in chimps and humans found that they were 99 percent identical; a similar comparison of proteins from two species of frogs turned up differences thirty to forty times greater. This suggests that humans and chimps may have diverged not twenty to thirty million years ago, as textbooks usually say, but just five or six million years ago (though there is evidence to the contrary in recent hominid-fossil finds).

Some years ago Dutch anthropologist Dr. Adriaan Kortlandt suggested an explanation for the surprisingly humanlike abilities of the chimpanzee. Chimps, he said, may have evolved, like our own ancestors, in open woodlands and on grassy plains, an environment that fosters hominid skills such as hunting and walking on two legs; but eventually the early humans invented weapons like the spear and drove the apes back into the jungle, where they lost many of those skills. According to Kortlandt's "dehumanization hypothesis," then, today's chimp is a kind of regressed hominid.

"The first time I heard that theory," Menzel told me, "I thought it was ridiculous. But the longer I've worked with chimps, the less ridiculous it has seemed. The truth is, we really don't know what they're capable of."

There is, I think, some danger that public opinion may swing too far in the other direction, that we may begin to expect too much of chimps—not the scientists who work with them and really know them, but the rest of us. In her book *Why Chimps Can Read*, Ann J. Premack notes that some of those who adopt a chimpanzee child eventually become disillusioned. When the animal's limitations finally become clear, they simply can't forgive it for not becoming human.

Judged by human standards, a home-raised chimp grows up to be something like a retarded child. But judged as another species, a creature in many ways ultimately mysterious and unknowable, it's a marvelous being indeed.

16.
The Wordless Signals Humans Send

Language is civilization itself.

—*Thomas Mann*

Language is a poor thing. You fill your lungs with wind and shake a little slit in your throat, and make mouths, and that shakes the air; and the air shakes a pair of little drums in my head . . . and my brain seizes your meaning in the rough. What a roundabout way and what a waste of time.

—*George du Maurier*

According to anthropologist Ray Birdwhistell, in any human conversation, no more than thirty-five percent of the social meaning is communicated in words. All the rest is nonverbal.

When I first heard that figure, I thought it had to be an exaggeration; but then I spent a year and a half researching a book on nonverbal communication, interviewing the scientists who were studying it. During that time some odd things happened to me, and I began to believe in Birdwhistell's percentage.

The strangest experience was one I had with Ray Birdwhistell himself. He's the inventor of the young science of *kinesics*, the study of body motion, and back in 1969, I spent about twelve hours with him while he gave me a crash course in his subject. A tall man with a deep and resonant voice, he bombarded me with new and sometimes outrageous ideas about everything from family resemblances to courtship. At one point, he asked if I found him hard to follow. It seems people sometimes complained that he had a habit of leaving his sentences unfinished and of jumping from one subject to another. I assured him that I didn't find him at all difficult to understand.

When I headed home with my pile of tape recordings, then, I was quite sure I knew what Professor Birdwhistell had said on any number of subjects. However, when I played the tapes back, it all began to seem rather muddy, and when I actually sat down and transcribed them, I panicked. Many of the things I was sure he had said simply weren't there—they had disappeared somehow into the gaps between unfinished sentences. In spite of that, I thought I knew in most cases how the sentence—even the paragraph—would have ended; and I was too embarrassed to call up and explain my problem. So I wrote a chapter of the book and, with considerable trepidation, sent it off to Birdwhistell before publication to get his reaction. To my incredulous relief, I discovered that it was accurate. But now I was even more puzzled than before: either I myself had an unsuspected talent for mind-reading or he had somehow managed to convey his thoughts—largely abstract ideas—nonverbally.

Months later another researcher finally suggested the real explanation. Professor Birdwhistell had simply watched me closely. As soon as he was sure (probably from my facial expression) that I understood what he was saying and where he was heading with it, he switched to another sentence or even another topic. So I *had* understood what I thought I'd understood, even if it wasn't on the tapes.

What seems significant about this in retrospect is that Birdwhistell apparently did, deliberately and with great skill, something the rest of us also frequently do—but on a level outside awareness. For humans constantly send and receive body signals, though we're seldom conscious of doing it, since we focus instead on the words that we exchange. We come away with a feeling that the other person understood us or didn't, liked or disliked us, was lying or telling the truth, but we can't really say where the feeling came from.

Fifteen or twenty years ago people scoffed at the idea that body movements could convey messages, and the half-dozen scientists who pioneered the study of nonverbal communication were treated with a kind of rejecting curiosity. Today, of course, Americans have come to accept the fact that there is such a thing as body language. However, it's a much misunderstood phenomenon, for most people believe it's simply a way we leak our secrets. A woman who sits with knees crossed and arms folded across her chest, for example, is supposedly signaling that she's unapproachable. Actually, body language is much more subtle and more complex than that.

It's also more ancient, for the truth is that we still send many of the same messages that animals send and in much the same way—without words; because when we aren't programming computers or producing television programs, we still have to deal with the same problems other animals face: we must compete, court, mate, raise our young—meet our fellow humans face to face.

Most nonverbal signals are not, of course, programmed in human genes. Instead, they're learned in childhood, absorbed automatically at the same time that we're learning the spoken language of our culture, so that as an adult, a Frenchman, for example, moves like a Frenchman, and flirts and asserts himself nonverbally using the signals Frenchmen use, while an Anglo-American moves and flirts and so on in the Anglo-American way. (Like some frogs, birds, and other animals, humans go in for a bewildering array of local dialects, both spoken and unspoken.)

Some years ago anthropologist Robin Fox suggested a plausible-sounding explanation for the way nonverbal signals work. He pointed out that because humans, as they shed their dependence on instinct, did still have the same relationship problems to solve, they had to develop learned behaviors that would work the way instincts do for the lower animals: signals that could be sent and received on a level outside of conscious awareness, so that no one had to take time out to think about them; that were sometimes automatic responses to situations; and that were common to all members of the population. Hence the whole repertoire of nonverbal behaviors. Because this repertoire is learned outside our awareness, it tends to persist from generation to generation, but—and Fox called this the element of genius—it can, if necessary, change within a generation, as innate behaviors never could.

Some of the people who study nonverbal communication actually call themselves human ethologists and are very aware of the animal parallels. Others start with an anthropological bias and are mainly interested in the cultural differences in the nonverbal code—in the fact, for example, that South Americans like to stand much closer together for conversation than North Americans do. Then there are psychologists, who design elaborate laboratory experiments, inducing subjects to lie, perhaps, and then examining the way this affects eye contact.

The kinesicists like Birdwhistell use an entirely different approach. They work with films, most often with films of psychotherapy sessions, for this most verbal of twentieth-century preoccupations has provided most of the raw material for the study of nonverbal communication. Using an ingenious notation system that Birdwhistell devised, the kinesicist may record every tiny movement of every participant in every frame of a segment of film—an incredibly painstaking and time-consuming task. Once the film has been transcribed into the Birdwhistellian symbols, the researcher goes over the transcription looking for patterns—for sequences of signals that are repeated.

To me, the most remarkable patterns the kinesicists have unearthed are the ones they dubbed *regulators*. For example, Dr. Albert Scheflen, a long-time colleague of Birdwhistell's, described the following sequence from a psychotherapy session in which a psychiatrist was seeing a family—mother, father, grandmother, and daughter—together for the first time. In the first step of an intricate exchange of signals, the mother would turn flirtatiously to the therapist. Resting a hand on her hip, she would extend her legs, delicately crossing her ankles, and, leaning forward, she would speak with great animation for perhaps twenty or thirty seconds. Then, quite suddenly, she would sink back in her chair again, withdrawing so completely that she looked almost autistic. When the film was run through in slow motion, it became clear what triggered her withdrawal. Each time she set out to charm the psychiatrist, her husband would start to jiggle one foot nervously. At this, both the daughter and the grandmother—who were sitting on either side of the mother—would cross their knees so that with their legs they boxed her in. Immediately afterward, she subsided into the chair. This sequence occurred eleven times in just twenty minutes of the film. It's not really a surprise, then, to learn that the mother's flirtatiousness was considered a family problem.

Almost certainly the family members weren't at all conscious of the body messages they were exchanging. Nevertheless, Dr. Scheflen believes that all families probably have sets of signals such as this that they use, all unaware, to regulate one another's behavior.

Another group of research films nicely illustrates the way regulators probably develop. The therapist-patient pair in this instance had their sessions recorded over a period of weeks. In the early films, the therapist often reminded the patient that he was supposed to say anything that came into his head; each time he delivered this admonition, the doctor also cleared his throat. After a few sessions, all he had to do was clear his throat and the patient would stop talking small talk and get to work recalling childhood memories. In one of the later sessions, the patient himself suddenly cleared his throat and broke off what he was saying to speak of his childhood. The process by which throat-clearing came to mean "get down to business" reminds me irresistibly of the way Emil Menzel's chimps gradually evolved signals for initiating tandem walking.

These two films also helped me understand one of the basic tenets of kinesics: that there will never be a dictionary of nonverbal signals, since no signal has a single, inevitable meaning (actually, few words do, either). To translate a behavior such as knee-crossing or throat-clearing, one must look at the whole context: at what the other parts of the body are doing, at how other people are behaving, at what is being said, at the entire situation.

As I said earlier, very few signals appear to be born into the human individual. Even those that seem as if they might plausibly have a genetic basis—because all humans use them and because in some cases they're also to be found among the other primates—can sometimes be equally well attributed to a kind of logic of the body. Take human greetings, for example: they bear a startling resemblance to those of the chimpanzee, for one chimp meeting another in the forest will often bow or crouch down in greeting or may actually hug and kiss the other or slap her on the back. This could indicate that, as Jane Goodall once put it, we may share with the chimpanzee "an ancestor in the dim and very distant past; an ancestor, moreover, who communicated with his kind by means of kissing and embracing, touching and patting and holding hands." Or it could simply be a logical way for one animal to reach out and reassure another when the animal has hands and arms and can get about, if it wants to, on two legs.

Another example is the threatening stare, a signal used and understood by many mammals. The alpha wolf, you will remember, has only to focus intently on a subordinate for the lesser animal to signal submission with drooping body and carefully averted eyes. Many kinds of monkeys and apes also threaten with a stare, and so do humans: imagine that you're riding on a subway or bus and look up to find that the man across from you has his eyes fixed intently on your face and is wearing an absolutely deadpan expression; even after you catch his eye, he doesn't look away. Most people would find this experience distinctly uncomfortable. Further evidence of the potency of the stare is the fact that the evil eye, the baleful look that can main or kill, has been feared throughout much of human history, right up to the present. Just a few years ago a department-store owner installed a frieze of watching eyes to try to discourage shoplifters; the ruse was surprisingly successful, and sales didn't suffer at all.

However, though the human response to the deadpan stare *may* be partly innate, it could also be, once again, simply the logic of the body, both with humans and with other animals. Where a creature looks shows what it's paying attention to. When it stares, then, with a deadpan expression that gives no clue to what its intentions are, that's almost bound to seem threatening.

Whatever the status of the stare, certain human facial expressions definitely appear to be coded in the genes, though presumably not in quite the same way as are the chemical signals of ants and the mating calls of frogs. Two lines of evidence suggest a hereditary base: cross-cultural studies have demonstrated that people all over the world smile with joy, scowl in anger, grimace with disgust, and so on; and, in addition, children who are born blind and so have no chance to observe and learn facial expressions develop all the usual ones. They cry, pout, wear typical expressions of anger, fear, and sadness, and produce their first social smiles at the age of five weeks, just as sighted babies do.

Ray Birdwhistell cautions, however, that the *meaning* of any particular facial expression is culturally prescribed, since it's our culture that decrees when it's appropriate to smile or cry or fly into a rage. The American South, for example, is a high-smile region—people there are expected to smile a lot—while New England and the Great Lakes region are much more dour. In fact, in Georgia, someone who doesn't smile very much is apt to be asked whether anything is wrong, while a southerner in Boston who smiles with southern frequency might actually be asked what's so funny. Unaware of this regional discrepancy, northern reporters during the 1976 Presidential campaign commented loudly and often on Jimmy Carter's perennial smile.

Birdwhistell believes that we even learn the faces we wear, which is why the people in a region often look so much alike even though they're not related. It seems babies are born with features that are splendidly soft and unformed. Gradually the eyebrows become set at a certain distance above the eyes and the scalp settles into position. The nasal septum rises between the ages of nine and eleven, and the mouth assumes its final shape still later, once all the permanent teeth are in. Absorbing the set of the faces they see around them, then, youngsters raised in Wisconsin tend to grow up with the remarkably smooth and unlined foreheads typical of the region. Similarly, New Englanders learn to be rather thin-lipped, while natives of western New York State project the lower lip slightly and carry it over the upper. Americans generally assume that they've inherited their looks. However, a few minutes in front of a mirror spent trying on a different way of holding the mouth and experimenting with various eyebrow levels will convince most people that the face they wear wasn't the only one possible for them.

It seems remarkable that human children should learn things like facial expressions—to say nothing of the other nonverbal signals—without being aware that they're learning anything at all; but it becomes plausible when you realize that humans are, on some level, enormously responsive not just to words but to other bodies. To begin to get a feel for this, however, you have to find a way to tune out that distracting verbal message. One way to do it is to turn on the television, leaving the sound off. Another is to quietly watch people who are out of earshot. Try to think of them as some alien species of animal—as well-dressed apes.

Imagine, for example, that two individuals, a man and a woman, meet in a public place. First they greet one another, perhaps by touching hands in an appropriately apelike manner. Among animals, a greeting is an appeasement ceremony. Anyone who doubts that it serves the same purpose for humans might try *not* greeting friends and acquaintances for a few days.

Deciding to stand and talk for a few minutes, our two humans take up positions a few feet apart. Like many other animals, humans need to preserve a certain distance between themselves and their conspecifics in order to feel comfortable. In this case, the distance chosen is a kind of dialect and reflects the shared culture of the pair: they stand a thoroughly Anglo-American three feet apart. If they had been Cuban, instead, two feet apart would have been ample. The distance also reflects their relationship, for if they were lovers rather than just friends, they would probably stand closer together.

The man faces the woman with his whole body and looks at her fairly steadily, indicating, in the way of animals, that she has his full attention. The woman, on the other hand, though the lower part of her body faces him squarely, has turned her upper torso aside, and she glances away much more often than he does. When a group of people approach and the woman is forced to step forward momentarily to get out of their way, she raises one arm up and across her body almost as if to shield herself, for she has moved in to a distance more intimate than the relationship warrants.

As the woman steps back again, the two are joined by a second male. It's instantly clear that he's of superior status, since he carries himself very much the way the alpha wolf does: after the initial greeting, he stands tall, with his head up, unsmiling; his gestures are lively and vigorous, his voice loud and decisive. Emil Menzel pointed out to me how surprising it is that humans and chimps, in asserting dominance, use the same simple-minded tactic that wolves and fish use—they draw themselves up so as to appear larger—though certainly even chimps are smart enough to see through such a subterfuge. Though the signal may be innate, in humans it might also be simply the wisdom of the playground lingering on into adult life, since any child knows that the bigger kids more often get their way. Statistics suggest that this still actually holds true in adult life: chief executives tend to be taller than the average.

Returning to our street-corner conversation, we see that the three people have settled now into particular postures. The woman and the first male are standing in an identical way, with their arms folded, while the second, dominant male has his hands in his pockets. When the first man drops his arms to his sides, the woman soon follows suit, almost as if she's imitating him. She is not, of course, deliberately doing any such thing, though in sharing his postures, she expresses a certain amount of rapport.

There are no courtship cues here—no preening and self-grooming, no prolonged, intimate eye contact—but all three are busy sending gender signals as they turn and begin to walk along the street together. The woman carries her pelvis tilted forward and up; she keeps her legs fairly close together and swings her arms from the elbows down. The two men, on the other hand, walk with their thighs somewhat apart, arms swinging from the shoulder, pelvis pushed back. Gender signals were first observed and described by Ray Birdwhistell, who reported that none of them are truly universal. Though people in seven very different cultures that he studied could describe ways of moving that were distinctively masculine or feminine, what was thought to be feminine in one culture was sometimes held to be masculine or neutral in another. For example, males in eastern Europe walk with their legs close together, as American women do, and in the Far East, men swing their arms from the elbows, not the shoulders.

Putting together the fact that every culture studied so far has gender signals and the fact that they differ from one culture to the next, Professor Birdwhistell concluded that humans must have a strong need for such signals, perhaps because as a species we're relatively unimorphic. There are, he pointed out, animals in which male and female look so different that it's hard to believe they belong to the same species, and others in which the sexes look so much alike that it's difficult to tell them apart—they are unimorphic. He places humans closer to the unimorphic end of the scale because there's so much overlap in all the distinguishing features we call secondary sexual characteristics: there are women with deep voices and men with high ones; women who are flat-chested and men who are not; bearded ladies and beardless men, and so on. We tell male from female, then, by adding up all these minor differences, and we also use the fact that men and women move differently and tend to dress differently, to help us make the distinction.

Body movement conveys a great deal, but humans also communicate in other ways that they're not generally aware of. There is smell, for example. Undoubtedly we make much less use of our noses than do many other mammals, but do we really use them as little as we think we do? Ten years ago, in a much-quoted article in the *New York State Journal of Medicine*, Dr. Harry Wiener wrote, "The fact is that our skin does contain a profusion of odor glands which rivals that of other animals. . . . They cover our body from head to toe; their structure is extremely complex; and there are so many individual types that complete anatomic classification has never been achieved."

Dogs can, of course, track a man given his scent from an article he has touched, indicating that every human has a kind of olfactory signature. Given the scent of progesterone, police dogs can pick out rods held by women who are pregnant or are in the second half of the menstrual cycle—both are times when the progesterone level goes up. It seems unlikely that all those human scent glands survived thousands of years of evolution for the benefit of dogs and other sharp-nosed animals, and Wiener suggested that humans, too, may respond to faint smells, but on a subconscious level.

However, it's easier to demonstrate the existence of human odor messages than it is to prove that humans respond to them. Experiments do suggest that a breast-fed baby may be able to recognize the smell of its own mother, and in one recent study, the majority of college students tested could tell by sniffing which of three T-shirts they themselves had worn and could also tell a shirt that had been worn by a male from one worn by a female. Such studies are suggestive, but the evidence for an olfactory subconscious is still sketchy, at best.

Besides body movements and smell, there is also paralanguage—the sound of the voice, apart from the words that are spoken. If you were to listen to a tape recording of people conversing in a language you didn't understand, you could probably guess whether they were male or female, their approximate ages, whether or not they were equals, whether they were friends or strangers, and perhaps something about the emotions they were experiencing, since anger, depression, happiness, and so on are all conveyed pretty clearly by what is loosely called tone of voice.

At this point, however, a word of caution seems in order: paralanguage can't really be separated from language—nor, for that matter, can body signals. It's sometimes suggested that the verbal system is used to convey information, while emotions are communicated nonverbally, but the dichotomy is not so clear-cut. To some extent, we perform our sentences as we speak them, using small-scale movements to enact the syntax of the message. When Americans ask a question, for example, at the end of the sentence, the voice rises, and at the same time the eyelids, head, hand, or some other body part usually lifts. At the end of a declarative sentence, the pitch of the voice drops—and so does some body part. We also use a complex set of nonverbal cues in conversation to let another person know that we're getting ready to yield the floor.

Then there is the odd phenomenon called interactional synchrony. It seems that humans almost literally "dance" to the rhythms of their own speech. As an individual talks, her gestures, hand and finger movements, head nods, eye blinks—whatever small motions she makes—synchronize with the stress points of her speech. And when she stops to listen to someone else, her body begins to dance instead to *his* beat. The basic tendency is easy to demonstrate: just ask a friend to beat out a rhythm with his finger and then talk to him. His taps will quickly begin to coincide with the stress points or the syllable divisions of your speech.

In real life, however, interactional synchrony is subtle and difficult to see, though it becomes obvious in films run in slow motion. A speaker's head begins to move to the right, and exactly at that instant—in that frame of the film—the listener's hand lifts. As the head stops and moves back to the left, the hand changes direction, too; the head movement speeds up and so does the hand. As Dr. William Condon, the discoverer of interactional synchrony, puts it, it looks as though both individuals were puppets moved by the same set of strings.

In films shot at twenty-four or forty-eight frames per second, the synchrony seems to be instantaneous, but when there are ninety-six frames for every second, a lag begins to show up between the words and the body motions, as if the sound were being processed very quickly at some lower neurological level. But sometimes even during silence, people move together, apparently reacting to visual cues in the absence of verbal ones.

Interactional synchrony seems to be universal: it's there in films of Eskimos, of African Bushmen, and of middle-class Americans. It's also there from the day of birth, for mother and baby lock in over long periods of time with a sharing of movement. Films of American babies just twelve hours old show that they will synchronize not only with the speech of someone in the room but also with a tape-recorded voice, whether it speaks English or Chinese. They don't dance in the same way with tapes of rhythmic tapping or of a human voice uttering isolated vowel sounds. Condon notes that if infants dance to the sound of the human voice from the beginning, then by the time they begin to speak, they have already laid down within themselves the form and structure of the language system of their culture.

However, curiously enough, humans aren't the only animals to move in sync. When Condon analyzed a film of a mother chimp grooming her baby, he found that they, too, moved to the same beat. And when he worked on a film of the chimp Viki—the one who could speak four words—he discovered that when her human foster father spoke, Viki moved in synchrony with his words, and she also "danced" to clicking sounds that she made herself.

Dr. Condon believes now that interactional synchrony is probably pananimal—that it exists even among the lower species. He points out that any organism has to be able to track what's going on in the world around it, to entrain with it and flow with it so that it can respond appropriately. The tracking takes place, he suggests, in the brain stem, and accounts for synchrony.

It should be obvious by now that there is a lot more to human communication than most people realize. In many conversations, although, ostensibly, information is exchanged, the more important transaction, as with animals, is the management of the relationship itself, and to a large extent, this is done nonverbally. In fact, Ray Birdwhistell believes that in any twenty-four-hour period, the amont of *new* information passing between any two individuals, except perhaps professional colleagues, probably doesn't exceed five minutes. "All the rest is just reassuring one another," he told me, "about who they are and how they feel about each other, about how the world is working and all the things that go into flowing with one another. Even if I were to talk to you about mathematics, we'd need to have constant reports on ourselves and the room beyond, and they'd be part of the message.

"Years ago I started with the question: how do body motions flesh out words? Now I ask instead: when is it appropriate to use words? They're very appropriate to teach or to talk on the telephone, but you and I are communicating on several levels now and on only one or two of them have words any relevance whatsoever. These days I put it another way: man is a multisensorial being. Occasionally he verbalizes."

17. Language— The "Pernicious" Gift?

Most people are determined to hold the line against animals. Grant them the ability to make linguistic reference and they will be putting in a claim for minds and souls. The whole phyletic scale will come trooping into Heaven demanding immortality for every tadpole and hippopotamus. Better be firm now and make it clear that man alone can use language and make reference. There is a qualitative difference of mentality separating us from animals.

—WORDS AND THINGS,
by Roger Brown

It's time now to come back to some of the questions that were raised at the beginning of this book. Specifically:

Can what the signing and reading apes do really be called language?

Is animal communication a completely different skill than human speech, or have we simply improved on what, as wordless mammals, we were once able to do?

What do the ape language projects mean, and what kind of practical impact will they have—along with all the other flourishing studies of animal communication—on the lives of animals and humans?

The answer to the first question seems to depend largely on whom you ask, since some linguists still insist that what the apes produce can't accurately be called language. The truth is, though, that we have no idea what the language limits are for apes. How much they can learn and how far they can go with syntax are still questions for the future. Probably it will always be possible to keep redefining the word "language" so that signing and reading apes can't quite qualify. However, looking at their accomplishments to date, it seems clear that this is already (to borrow Roger Fouts's analogy) a bit like insisting that a Volkswagen isn't a car because it's not a Cadillac and can't do all the things a Cadillac can do.

To turn to the second question: over the years, linguists, trying to work out exactly how language is different from animal signaling, have compiled lists of special "design features" that are supposed to distinguish language, to prove that it's different in kind and not just in degree from what animals do. Some lists of these features have run to well over a dozen items; however, eight seem to be particularly central to what language is. They are: semanticity, arbitrariness, cultural transmission, duality, displacement, prevarication, structure-dependence, and creativity.

In plain English, *semanticity* refers to the fact that languages use symbols (usually, words) to stand for objects and actions. We know now that this isn't beyond the capabilities of at least some species of animals. The honey bee dancing the location of a food find uses symbols, as does the chicken when she squawks "rehh" for danger from the air and cackles "gogogocock" for a menace on the ground. In both cases, the symbols are inborn; however, chimps both in the wild and in captivity seem to to be capable of agreeing on a meaning for particular gestures; this is closer to what humans do.

Arbitrariness simply means that the symbols humans use generally bear no resemblance to the objects or actions they represent. With a few exceptions, such as the word "buzz," which sounds like the sound it names, the choice of any particular word to label something is entirely arbitrary: a "cat" could as easily be called a "dobo." Here, again, there are some animal signals that seem to qualify. Colin Beer's gulls, for instance, vary their long calls to indicate that they are addressed to their own chicks and not to any other bird.

The third feature is *cultural transmission*—meaning that human languages are learned, passed down from one generation to the next. As we have seen, there is cultural transmission to some extent among the birds, frogs, and so on that sing or croak in dialect.

Languages also have *duality*: the phonemes or bits of sound from which words are constructed are almost always meaningless in themselves, though they create something meaningful when they're combined. Similarly, most of the letters of the alphabet have no meaning as long as they stand alone.

When we turn to animal signaling, it's obvious that the individual notes of a bird's song have no meaning in and of themselves, though when they're combined, they do convey a message. The laughing gull provides an even more telling example, since from a small set of call notes, the bird constructs a repertoire of about a dozen different calls. According to Colin Beer, the notes themselves are comparable to phonemes—they're minimum units of sound—and the calls operate, as words do, as minimum units of sense.

Another vital feature of language is *displacement*: humans aren't chained to the present but can talk about things distant in time and place. Of course, the honey bee dancing the location of food some distance away does this, too—though some critics insist that she may simply be rehearsing her next flight rather than referring to a particular location. The chimpanzee mother who coped with her dawdling daughter by picking a grass stem and moving toward a specific path might also have been communicating about something distant (a termite mound), though we can't really be sure that she was giving a signal rather than unintentionally providing information as she went about her business.

Prevarication—the ability to lie—is also considered a special feature of language. But even without language, Bruno lied to Booee when he gave an alarm bark to get the hose away from him, while among Emil Menzel's chimps, Belle tried hard to lie to Rock. Menzel believes that, whether animals lie or not, they often withhold information—which, as any small child knows, pragmatically can come to much the same thing. A young rhesus monkey, eager to get hold of a plastic banana but inhibited by the presence of an older monkey, will sometimes act as if he doesn't even see the banana until the older animal leaves—at which point he'll race over and grab it.

Then there is *structure-dependence*. Language isn't just words strung together, it's highly patterned; and clumps of words that belong together—a noun or verb phrase, for example—can be slotted into a sentence in different positions without really changing the meaning. "The green crocodile swallowed the pocket watch" is essentially the same as "The pocket watch was swallowed by the green crocodile" and is related to the question "Did the green crocodile swallow the pocket watch?" Though I've found no reports of structure-dependence in animal signaling, I can't help wondering whether humans would recognize it if it were there.

Creativity is the last of the design features I want to touch on. Humans don't simply learn set phrases; instead, thanks to the way syntax operates on vocabulary, they can combine words to talk about anything—there is no limit to the number of different sentences they can construct. The fact that the language-project apes, as soon as they've been taught three or four signs or symbols, begin to combine them in new ways suggests that apes, too, are capable of some degree of linguistic creativity—unless one assumes that in giving them language, we somehow made them "creative," which seems unlikely. We don't really know whether apes in the wild have a signaling system capable of creativity.

By now, it should be clear that there are many surprising similarities between animal signaling and human language. Most, if not all, of the design features of language can also be found in the signaling system of some species somewhere. Though no system seems to have all of these features and none has the power of language, the continuities are there. Surely we can learn from them, and surely we're being unnecessarily defensive if we doggedly maintain that in language, even if in nothing else, humans are unique.

However, I don't want to overstate the case. Humans *are* specialized for language, the way birds are specialized for flight. The human tongue, teeth, vocal cords, and brain are all preadapted for speech. In fact, talking requires such complex coordination of so many different muscles that it seems we must be genetically preprogrammed to some extent to be able to do it at all. Obviously the acquisition of language also represented an incalculable leap forward for the human species—and it may indeed be primarily responsible for what we call civilization.

To turn now to the third question, I think the impact of the recent discoveries about animal communication is already being felt in two very different ways.

First of all, in the academic world, glottogenesis is suddenly a hot topic again, and psychologists and linguists have been eagerly debating how humans came to have language. It's a subject that has long intrigued scientists, and numerous theories have been advanced, but until the chimp language projects, they all seemed purely speculative. Now, suddenly, the gestural-origin hypothesis advanced some years ago by anthropologist Gordon Hewes looks decidedly plausible.

Hewes has argued that speech is a relatively recent development in the long history of the human race, and that before there was speech, there were gestural languages. He bases his theory partly on the fact that Neanderthal men and women apparently didn't have a vocal tract capable of all the articulations of modern speech, though they hunted, made tools, and buried their dead. Some anthropologists believe that a creature totally devoid of language would have been incapable of hunting or tool-making and unlikely to care about burial. In addition, the recent successes of the signing apes suggest that the early hominids probably had the mental ability necessary to invent language even before they developed the physical ability to speak. Furthermore, Hewes points out that deaf children left to their own devices will often spontaneously develop their own shared system of signs.

Dr. Adam Kendon, who is a psychologist and kinesicist currently on a research fellowship at the Australian National University in Canberra, has come up with evidence for the gestural-origin theory that's even more intriguing. Proceeding as kinesicists do with a frame-by-frame analysis of films of human conversations, Kendon discovered that people often enact with movements the ideas they're uttering, even as they express them in words. For instance, in one film, as a woman gropes for the word "cake," she uses her index finger to draw a circular cake shape in the air. Speaking of the way someone "jumped up," she flings her own arm up. Gesticulations such as these often begin *before* the matching spoken phrase begins, and in fact sometimes, as in the "cake" example, even as the person hesitates, searching for the right word, she is already making the appropriate gesture. To Kendon, this suggests that gestures reflect an early stage in the process of translating an idea in the mind into an idea expressed to another individual.

The average person doesn't give much thought to the process of utterance—to the way ideas get out. Linguists, however, have several conflicting theories about it, and one fits nicely with Kendon's data. It holds that even as we utter one clause, we are already picking out in our heads one or more key words from the next. At first, though, the key word is more of a "word idea"—a kind of category; it might come out either "cabinet" or "cupboard," "shirt" or "blouse," for example. Once we've hit on the specific word and begin to utter its clause, we quickly slot in the rest of the words with their appropriate endings in the places where they belong. The evidence that supports this theory comes largely from studies of slips of the tongue and is too involved to go into here; the real point is Kendon's contention that sometime very early in this process of utterance, perhaps just as the key word or word-idea is being selected, the idea may travel to the hands.

Kendon's observations indicate, then, that ideas are more swiftly translated into gestures than into words—there seem to be fewer steps involved in the process—and that the gestural channel is actually easier to use. The fact that children tend to produce their first signs long before the age at which speech typically develops further supports his contention. As Kendon himself put it, writing on the gestural origin theory: "The least we can say about this is that we would not, perhaps, expect a more elaborate and time-consuming method of utterance to be the one that was first developed in language evolution."

Linguists like to speculate not only about how language evolved but about why. What formidable environmental pressures forced the development of human linguistic abilities? Some suggest that hunting made language necessary, but to me this isn't entirely convincing, since wolves, killer whales, and other animals hunt with great skill, presumably without benefit of anything close to human language. Others believe that humans needed speech for tool making. However, verbal instructions aren't really the most efficient way to teach motor skills. Archaeologist Desmond Clarke of the University of California at Berkeley teaches a course in the making of Stone Age tools. They're surprisingly difficult to make without more advanced tools, such as hammers and chisels, to help, and Professor Sherwood Washburn of Berkeley notes that though Clarke delivers verbal instructions, the students watch his hands most of the time.

So the pressures that fostered language remain a mystery. Nor do we know, assuming that language began as a gestural system, what forced the switch from signing to speech; but perhaps in the future, research on animal communication will suggest answers to these questions.

In the meantime, the ape language projects have generated an entirely practical payoff of a very different kind. Critics in Congress and outside it who once complained that thousands of dollars were being wasted on a whim—to find out whether it was possible to teach an ape to talk—were effectively silenced when it became known that both the artificial languages designed for apes have been adapted and are being used, with stunning success, to teach language to children who are unable to talk. In fact, once they have begun to master a plastic language like Sarah's, many children are able to graduate to the spoken word.

Because of a brain tumor, my own mother was aphasic—more or less speechless—for a year before she died, able apparently to understand what was said to her but unable to manage much in the way of intelligible speech. She would produce a string of words with no real syntax, perhaps one or two of which had to do with what she was trying to say, and we would play a guessing game. "Are you too hot?" I'd ask her, or "Is it time for your lunch?" She could say yes and no, though sometimes she said one when she meant the other; but once the word was out, she could hear it and knew if it was wrong. When that happened, she would shake her head in frustration, so in the end we got the answer, anyway.

My mother's disability haunted me for years, for it made me see how imprisoned we are—how desperate—without speech, without an outlet for our feelings, for love or for fear. And I wondered what it was like inside her head. Could she think without words? Where did the speech process break down? Were the thoughts—the words—logical and organized while they were still in her mind, and were they somehow tripped up on the way to her tongue? Or was her thinking a kind of muddled word stew, like her speech?

Now I've begun to believe that perhaps neither answer is right; for in the course of writing this book, I've discovered that we are, many of us, prisoners of language, unable even to imagine what it would be like to think without words. I couldn't imagine it either until one night a few months ago, when I was mulling over the Saturday-morning errands I had to run the next day. Suddenly I realized that I could plan my itinerary in an entirely visual way: I could visualize the man behind the counter at the hardware store, the sign over the door at the delicatessen, the inside of the cleaner's shop, plotting the order in which I'd run my errands, perhaps, in much the same way that Menzel's chimps planned the route they'd use to collect the eighteen piles of food he'd hidden. I noticed, too, that I didn't have to visualize every step of the route or even any step of it, and that for every store, I had a kind of visual symbol—a familiar human face, a mental image of the counter where I'd find flashlight batteries or whatever. Obviously one can use imagery to plan for the future, to recall the past, even to symbolize an abstract idea, since there is always some event in the past that stands in a very personal way for "love," for example, or even for "justice."

This doesn't, of course, mean that aphasics necessarily think in visual images or that that's the way animals do it; a dog might actually think in smells. But it does suggest that perhaps we shouldn't be quite so wedded to the idea that thinking *is* words and that it's impossible to think without them.

Which brings me to the deeper significance of recent animal-communication research and to the impact I believe it will have in the future.

Many scientists either deny that animals are conscious and self-aware—have mental experiences—or insist that whether they do or not is irrelevant, since there's no way to study animal consciousness scientifically. In 1976, then, when Professor Donald Griffin's book *The Question of Animal Awareness* was published, it caused quite a stir in some circles. Keeping his definitions simple, Griffin wrote that awareness is "the whole set of interrelated mental images of the flow of events"—events that may be taking place in the immediate situation or may be remote in time or place. He went on to say that "mental images obviously vary widely in the fidelity with which they represent the actual surrounding universe, but they exist in some form for any conscious organism."

Though Griffin doesn't suggest that the mental experiences of animals are necessarily anything like those of humans, he does suggest that it's time to try to find out what they *are* like. Seventy years ago, reacting to a general tendency to ascribe human feelings to a wide variety of animals purely because of anecdotal evidence, the early behaviorists laid down a working strategy: scientists were to assume that animals have neither mental states nor subjective experiences and to see how much of animal behavior they could explain on this basis. Originally it was, as Griffin explains, an agnostic position: it didn't actually deny that animals could think; it simply held that it was unproductive to try to analyze what went on in their heads.

Over the years, however, hard-line behaviorists came to talk as if animals were simply soft machines, responding automatically, like Pavlov's salivating dog, to external stimuli or to internal drives. Eventually ethologists told another version of the same story: to many of them, animals were soft machines responding automatically as they'd been programmed to do by their genes. For example, Griffin writes that it has been the "curious custom of most ethologists to stop short of interpreting an intention movement as evidence that the animal has a *conscious* intention." Since both humans and other animals can generally guess what a creature will do by its intention movements, it seems odd to assume that the animal itself isn't also anticipating its own future actions. For example, certain birds, when a predator threatens their nests, will pretend to be injured, flopping about on the ground as if they have a broken wing and all the while luring the predator away from their young. Some scientists have gone to great lengths to consider how this particular display, which presumably is genetically programmed, might have evolved, while denying that the bird acts with any conscious intention. Yet as Professor Griffin pointed out, it's perfectly possible that the display is indeed genetically programmed and that the bird still deliberately uses it to protect its young.

Griffin concludes that "the narrowing gap between animal communication and human language calls into question . . . the assumption that animals respond mechanically to external or internal stimulation, whereas humans speak with conscious understanding and intent." He believes that animal-communication studies will eventually provide a way to examine, objectively and scientifically, the mental experiences of other species, that they can supply a kind of window into the minds of animals.

Reading Professor Griffin's book, it occurred to me that those whose profession requires them to experiment on animals probably find it more comfortable to think of them as machines; then their needs and feelings needn't be taken into account. If their inner awareness is at all similar to ours, some sticky ethical questions are raised. Dr. Griffin mentions this fact in passing in his book, and he explained to me in an interview that he didn't really go into the ethical implications because he doesn't believe that scientists are any better qualified to comment on such questions than anyone else is. However, he also doesn't believe in not asking the questions just because the answers may turn out to be uncomfortable.

For me, though, over the time I have spent researching animal communication, the ethical issue has loomed larger and larger. It all came to a head one Sunday when I picked up the *New York Times Magazine* to find that it featured a story about Koko, Penny Patterson's signing gorilla. According to the writer, Harold T. P. Hayes, the San Francisco zoo had given Penny ninety days to raise $12,500 to buy Koko. If she couldn't come up with the money then, the zoo would take Koko back and use her for breeding. Female gorillas with breeding potential are rare in captivity, and according to the zoo's figuring, Koko was actually worth about $20,000.

Yet what the zoo proposed doing with Koko was, to me, unthinkable. At the age of six, she has an I.Q. between 80 and 90, the equivalent of a five-year-old human child; there are lots of adults functioning in human society who would test no higher than that. She also has a proven vocabulary of 300 signs and probably knows about 450 in all.

If Koko were returned to the zoo, it would mean the end of one of the most promising of all the ape language projects; but what's more significant, it might also mean the end of Koko. As Penny said, "To take her away from her family, her environment, to throw her in a zoo cage with a bunch of gorillas—it could kill her." For not only is Koko extremely attached to Penny, who is, for all practical purposes, her mother, but she probably doesn't know she's a gorilla. Raised among humans, when she was first introduced to Penny's new infant gorilla, Michael, she avoided direct contact with him. When Penny finally asked her what she was afraid of, she signed back, "Afraid alligator."

One way or another, Penny finally managed to raise the funds she needed to buy Koko, though as of this writing, she's still in debt because she had to borrow money when she bought Michael. In the meanwhile, though, the *Times* article raised some preplexing, fascinating, and important legal and ethical equestions.

Professor Theodore Sager Meth, a Seton Hall law professor who teaches a course in animal rights, believes that if Penny had not been able to buy Koko, a court might still have decided that the gorilla had a legal right to remain where she was. "The gorilla doesn't exist any more," he said in an interview with the *Times*. "Under normal circumstances, the only thing this animal doesn't have that we do is language. Now you have changed it. When you give it the conceptual apparatus for conscious reasoning, for mobilizing thought, you have radically altered it. You have given it the pernicious gift of language. If it has never been one before, it is an individual now. It has the apparatus for the beginning of a historical sense, for the contemplation of self." Since that's the case, Meth argued, then Koko's right to remain with the people she has known might well take legal precedence over the zoo's claim to property rights.

What bothers me about this reasoning is that it assumes we have created a different kind of creature by giving an ape language—which may be overrating language, though certainly we create a different kind of creature when we raise an ape as if it were a human being. Perhaps the argument should be that since apes are capable of learning language, they are enough like humans already to deserve to have at least some human rights before the law; and this would apply to all apes, not just to those who have been tutored in human methods of communication. However, as Professor Meth pointed out when I talked to him, the law is concerned not with what is just but with what is practical, and one of the practical, legal problems in regard to animal rights is, where does one draw the line? If signing apes have certain rights, what about nonsigners? And what about killer whales and wolves? Do ants have a right to raid my food supply or do I have a right to wipe them out in defense of my groceries? Where *does* one draw the line? It's easy to reduce the whole subject to absurdities, but I think the question itself should be asked, and probably will be asked, over and over again within the next few years.

For myself, what I've learned about animal communication has changed the way I think about animals. It has, in fact, left me rather uncomfortable. I can't see them any longer as creatures totally different from myself, experiencing existence in some unimaginable way in the silence of their own skulls, responding in knee-jerk fashion to the push of their own instincts or to cues from the outside world. Frogs, sparrows, and marmosets are very different from human beings, of course, and it would be foolish to think they're not; but they're also more like us than we've been willing to admit.

The truth is that we find it difficult to see animals objectively. Either we become terribly sentimental about them, endowing them with all the best human qualities—insisting that they are somehow more noble, more gentle, perhaps even more intelligent in their way than humans are; or we identify in them our own worst qualities, creating myths about savage predators and brutal, merciless killers. We eat animals, hunt them for sport, experiment on them; we also turn them into coddled pets and treat them like surrogate children. We are enormously ambivalent about them.

In the end, what I have come up with are a rash of uncomfortable questions and no answers. I'm not ready to become a vegetarian, nor am I sure I morally should. One must eat something, and many animals eat other animals—it's part of the balance evolution worked out. Nor would I want to see the cure for cancer, for example, delayed by even a few years in order to save the lives of experimental animals. It may be human chauvinism, but I value human lives above those of other species. It does seem true, however, that humans don't often stop to ask themselves whether the outcome of the experiment is worth the price in animal lives or animal pain. The answer to such a question may not come easily; nevertheless, it seems time it was asked more often.

Law professor Ted Meth says that asking the question *is* the moral issue; that ethics is the study of ambiguities, and that if there were no ambiguities, there would be no need for ethics, or their enforcement through law. If one eats meat with the full and not-so-comfortable awareness that this was once a living creature, and one eats it, perhaps, less often because of that, then that's a beginning and perhaps it's even sufficient.

There is today a growing concern for animals and for the environment. Meth believes that though this concern may ultimately benefit humans, it may already be too late for it to do animals any good. Individual humans may become better people as they become aware of the continuity of life, of both the precious differences and the similarities between animals and humans; but the expansion of the human species is overwhelming all other species.

Humans have such terrible power, Meth said, and we exist in such terrible numbers. We encircle the few enclaves in which wild animals still live, and every year we press in on them farther; and of course the pollutants we produce also take their toll. Gorillas are already a doomed species. In the future, all the apes left alive may well live in captive colonies in California or Florida or wherever, and the same may be true of wolves, dolphins, and other animals.

It's not a future I find pleasant to contemplate. But perhaps as we learn more about animals and as we begin to be able to communicate with them in their own terms, and sometimes even in ours, attitudes will change.

Perhaps if we can somehow reach into the minds of other species, more of us will come to sense the connectedness of all life. Perhaps then we will finally begin to exercise, on behalf of all the creatures of the planet, the foresight that language supposedly makes possible.

Perhaps it's not, after all, too late.

Bibliography

Chapter 2.

Of general interest—very relevant, very readable, or both:

Gardner, Beatrice T. and R. Allen. "Comparing the Early Utterances of Child and Chimpanzee." In *Minnesota Symposia on Child Psychology* 8: 3–24. Minneapolis: University of Minnesota Press, 1974.

Further references:

Beach, Frank A. "Beasts Before the Bar." In *Ants, Indians and Little Dinosaurs,* edited by Alan Ternes. New York: Scribner's, 1975.

Brown, Roger. *A First Language.* Cambridge, Mass.: Harvard University Press, 1973.

Farb, Peter. *Word Play.* New York: Knopf, 1974.

Gardner, Beatrice T. and R. Allen. "Two-Way Communication with an Infant Chimpanzee." In *Behavior of Non-Human Primates,* edited by Allan M. Shrier and Fred Stollnitz. New York: Academic Press, 1971.

Gardner, Allen and Beatrice. "Teaching Sign Language to a Chimpanzee, Washoe." *Bulletin d'Audio-Phonologie,* 4 (1974).

Linden, Eugene. *Apes, Men, and Language.* New York: Saturday Review Press, 1974.

Chapter 3.

Of general interest:

Fouts, Roger S. "Talking with Chimpanzees." *Science Year* (1973), 34–49.

Further references:

Fleming, Joyce Dudney. "Field Report. The State of the Apes." *Psychology Today,* Jan., 1974, 31–50.

Fouts, Roger S. "Language: Origins, Definitions and Chimpanzees." *Journal of Human Evolution* 3 (1974): 475–482.

Fouts, R. S. "Communication with Chimpanzees." In *Hominisation und Verhalten,* edited by Kurth and Eibl-Eibesfeldt. Stuttgart: Gustav Fischer Verlag, 1975.

Temerlin, Maurice K., *Lucy: Growing up Human.* Palo Alto, Calif.: Science and Behavior Books, 1975.

Chapter 4.

References:

Gardner, R. Allen and Beatrice T. "Teaching Sign Language to a Chimpanzee." *Science* 165 (Feb. 28, 1975): 664–672.

Chapter 5:

Of general interest:

Premack, Ann J. *Why Chimps Can Read.* New York: Harper & Row, 1976.

Premack, David. *Intelligence in Ape and Man.* New York: Halsted Press, 1976.

Further references:

Anders, Frank. "If You Can Teach an Ape to Read, Can You Do Something for My Retarded Child?" *New York Times Magazine,* June 1, 1975: 14–59.

Premack, David. "Language in Chimpanzee?" *Science* (May 21, 1971): 808–822.

Chapter 6.

Of general interest:

Rumbaugh, Duane M., ed. *Language Learning by a Chimpanzee.* New York: Academic Press, 1977.

Chapter 7.

Of general interest:

Bonner, John Tyler. "Hormones in Social Amoebae and Mammals." *Scientific American* 220 (June, 1969): 78–91.

Further references:

Bonner, John T. *The Cellular Slime Molds.* Princeton: Princeton University Press, 1959, 1967. Available from University Microfilms International, Ann Arbor, Michigan.

Bonner, John Tyler. "Some Aspects of Chemotaxis Using the Cellular Slime Molds as an Example." *Mycologia* (May–June, 1977).

Chapter 8.

References:

Barlow, George W. "Contrasts in Social Behavior between Central American Cichlid Fishes and Coral-reef Surgeon Fishes." *American Zoologist* 14 (1974): 9–34.

Hopkins, Carl D. "Sex Differences in Electric Signaling in an Electric Fish." *Science* 176 (June 2, 1972): 1035–1037.

McKaye, Kenneth R., and Barlow, George W. "Competition Between Color Morphs of the Midas Cichlid, *Cichlasoma citrinellum*, in Lake Jiloa, Nicaragua." In *Investigations of the Ichthyofauna of Nicaraguan Lakes*, edited by T. B. Thorson. Lincoln: School of Life Sciences, University of Nebraska, 1976.

McKaye, Kenneth, and Barlow, George W. "Chemical Recognition of Young by the Midas Cichlid, *Cichlasoma citrinellum.*" *Copeia* (May 17, 1976): 276–282.

Noakes, David L. G., and Barlow, George W. "Ontogeny of Parent-Contacting in Young *Cichlasoma citrinellum* (Pisces, Cichlidae)." *Behaviour* 46 (1973): 3–4.

Chapter 9.

Of general interest:

Topoff, Howard R. "The Social Behavior of Army Ants." *Scientific American* 227 (Nov., 1972): 71–79.

Schneirla, T. C. *Army Ants. A Study in Social Organization*, edited by Howard R. Topoff. San Francisco: W. H. Freeman, 1971.

Further references:

Topoff, Howard. "Behavioral Changes in the Army Ant *Neivamyrmex nigrescens* during the Nomadic and Statary Phases." *Journal of the New York Entomological Society* 83 (March, 1975): 38–48.

Wilson, Edward O. *Sociobiology*. Cambridge, Mass.: Harvard University Press, 1975.

Wolff, Anthony. "Building a Better Bug Trap." *New York Times Magazine*, Nov. 28, 1976: 38–58.

Chapter 10:

Of general interest:

Gould, James L. "The Dance-Language Controversy." *Quarterly Review of Biology* 51 (June 1976): 211–244.

Further references:

Gould, James L. "Honey Bee Recruitment: The Dance-Language Controversy." *Science* 189 (Aug. 29, 1975): 685–693.

Gould, James L. "Communication of Distance Information by Honey Bees." *Journal of Comparative Physiology* 104 (1975): 161–173.

Griffin, Donald. *The Question of Animal Awareness*. New York: Rockefeller University Press, 1976.

von Frisch, Karl. *The Dancing Bees*. New York: Harcourt, 1953.

Wenner, Adrian M. "Honey Bees." In *Animal Communication*, edited by Thomas Sebeok. Bloomington: Indiana University Press, 1968.

Chapter 11.

References:

Capranica, Robert R. "Vocal Response of the Bullfrog to Natural and Synthetic Mating Calls." *Journal of the Acoustical Society of America* 40 (Nov., 1966): 1131–1139.

Capranica, Robert R. "The Vocal Repertoire of the Bullfrog (*Rana Catesbeiana*)." *Behaviour* 31 (1968): 3–4.

Capranica, Robert R., Frischkopf, Lawrence S., and Nevo, Eviatar. "Encoding of Geographic Dialects in the Auditory System of the Cricket Frog." *Science* 182 (Dec. 21, 1973): 1272–1275.

Feng, A. S., Narins, P. M., and Capranica, R. R. "Three Populations of Primary Auditory Fibers in the Bullfrog (*Rana catesbeiana*): Their Peripheral Origins and Frequency Sensitivities." *Journal of Comparative Physiology* 100 (1975): 221–229.

Frishkopf, L. S., Capranica, R. R., and Goldstein, M. H., Jr., "Neural Coding in the Bullfrog's Auditory System—a Teleological Approach." *Proceedings of the Institute of Electrical and Electronics Engineers* (June, 1968): 969–980.

Hoy, R., Hahn, J., and Paul, R., "Hybrid Cricket Auditory Behavior: Evidence for Genetic Coupling in Animal Communication." *Science* 195 (1977): 82–84.

Hoy, R. R., and Paul, R. C., "Genetic Control of Song Specificity in Crickets." *Science* 180 (1973): 82–83.

Narins, Peter M., and Capranica, Robert R., "Sexual Differences in the Auditory System of the Tree Frog *Eleutherodactylus coqui*," *Science* 192 (April 23, 1976): 378–380.

Chapter 12.

Of general interest:

Marler, Peter. "Animal Communication." In *Nonverbal Communication*, edited by L. Krames, P. Pliner, and T. Alloway. New York: Plenum Press, 1974.

Beer, Colin. "Some Complexities in the Communication Behavior of Gulls." In *Origins and Evolution of Language and Speech*, edited by S. R. Harnad, H. D. Steklis, and J. Lancaster. New York: Annals of the New York Academy of Sciences, 1976.

Further references:

Beer, C. G.. "Multiple Functions and Gull Displays." In *Function and Evolution in Behaviour*, edited by Gerard Baerends, Colin Beer, and Aubrey Manning. Oxford: Clarendon Press, 1975.

Ford, Barbara. *How Birds Learn to Sing.* New York: Julian Messner, 1975.

Kroodsma, Donald E. "Reproductive Development in a Female Songbird: Differential Stimulation by Quality of Male Song." *Science* 192 (May 7, 1976): 574–575.

Marler, Peter. "Learning, Genetics and Communication." *Social Research* 40 (Summer, 1973): 293–310.

Marler, Peter. "On Strategies of Behavioral Development." In *Function and Evolution of Behaviour*, edited by Gerard Baerends, Colin Beer, and Aubrey Manning. Oxford: Clarendon Press, 1975.

Marler, Peter. "On the Origin of Speech from Animal Sounds." In *The Role of Speech in Language*, edited by J. F. Kavanagh and J. Cutting. Cambridge, Mass.: M. I. T. Press, 1975.

Marler, Peter. "Sensory Templates in Species-Specific Behavior." In *Simpler Networks and Behavior*, edited by John C. Fentress. Sunderland, Mass.: Sinauer Associates, 1976.

Marler, Peter, and Mundinger, Paul. "Vocal Learning in Birds." In *Ontogeny of Vertebrate Behavior*, edited by Howard Moltz. New York: Academic Press, 1971.

Chapter 13.

Of general interest:

Wood, Forrest G. *Marine Mammals and Man.* Washington, D.C.: Robert B. Luce, 1973.

Further references:

Cummings, William C., Thompson, Paul O., and Cook, Richard. "Underwater Sounds of Migrating Gray Whales, *Eschrichtius glaucus* (Cope)." *Journal of the Acoustical Society of America* 44 (Nov., 1968): 1278–1281.

Cummings, William C., and Thompson, Paul O. "Gray Whales, *Eschrichtius robustus,* Avoid the Underwater Sounds of Killer Whales, *Orcinus orca.*" *Fishery Bulletin* 69 (1971): 525–530.

Cummings, William C., Fish, James F., and Thompson, Paul O. "Sound Production and Other Behavior of Southern Right Whales, *Eubalena glacialis.*" *San Diego Society of Natural History Transactions* 17 (March 15, 1972): 1–14.

Evans, William E., and Bastian, Jarvis. "Marine Mammal Communication: Social and Ecological Factors." In *The Biology of Marine Mammals,* edited by H. T. Andersen. New York: Academic Press, 1969.

Fish, James F., and Vania, John S. "Killer Whale, *Orcinus orca,* Sounds Repel White Whales, *Delphinapterus leucas.*" *Fishery Bulletin* 69 (1971): 531–535.

Lang, T. G., and Smith, H. A. P. "Communication between Dolphins in Separate Tanks by Way of an Electronic Acoustic Link." *Science* 150 (Dec. 31, 1965): 1839–1844.

Norris, Kenneth S. *The Porpoise Watcher.* New York: Norton, 1974.

Chapter 14.

Of general interest:

Fox, Michael W. *Understanding Your Dog.* New York: Coward, McCann & Geoghegan, 1972.

Further references:

Fox, M. W., ed. *The Wild Canids.* New York: Van Nostrand Reinhold, 1975.

Chapter 15.

Of general interest:

Menzel, Emil W., and Johnson, Marcia K. "Communication and Cognitive Organization in Humans and Other Animals." In *Origins and Evolution of Language and Speech.* New York: Annals of the New York Academy of Sciences, 1976.

Further references:

Menzel, E. W., Jr., "Leadership and Communication in Young Chimpanzees." *Symposia of the Fourth International Congress of Primatology* 1 (1973): 192–225.

Menzel, Emil. "Natural Language of Young Chimpanzees." *New Scientist* 65 (Jan. 16, 1975): 127–130.

Teleki, G. "Chimpanzee Subsistence Technology: Materials and Skills." *Journal of Human Evolution* 3 (1974): 575–594.

Chapter 16.

Of general interest:

Birdwhistell, Ray L. *Kinesics and Context.* Philadelphia: University of Pennsylvania Press, 1970.

Davis, Flora. *Inside Intuition. What We Know about Nonverbal Communication.* New York: McGraw-Hill, 1971, 1972, 1973.

Further references:

Condon, W. S., and Ogston, W. D. "A Method of Studying Animal Behavior." *Journal of Auditory Research* 7 (1967): 359–365.

Condon, William S., and Sander, Louis W. "Synchrony Demonstrated between Movements of the Neonate and Adult Speech." *Child Development* 45 (1974): 456–462.

Coss, Richard G. "Reflections on the Evil Eye." *Human Behavior* 3 (Oct., 1974): 16–22.

Fox, Robin. *Encounter with Anthropology.* New York: Dell, 1975.

Scheflen, Albert E. "Communication and Regulation in Psychotherapy." *Psychiatry* 26 (May, 1963): 126–136.

Scheflen, Albert E. *How Behavior Means.* Garden City: Anchor, 1974.

van Lawick-Goodall, Jane. *In the Shadow of Man.* Boston: Houghton Mifflin, 1971.

Wiener, Harry. "External Chemical Messengers: 1. Emission and Reception in Man." *New York State Journal of Medicine* 66 (1966): 3153–3170.

Chapter 17.

References:

Aitchison, Jean. *The Articulate Mammal.* New York: Universe Books, 1976.

Griffin, Donald. *The Question of Animal Awareness.* New York: Rockefeller University Press, 1976.

Hayes, Harold T. P. "The Pursuit of Reason." *New York Times Magazine*, June 12, 1977, 21–79.

Kendon, Adam. "Gesticulation, Speech and the Gesture Theory of Language Origins." *Sign Language Studies* 9 (Winter 1975).

Index